SpringerBriefs in Economics

SpringerBriefs present concise summaries of cutting-edge research and practical applications across a wide spectrum of fields. Featuring compact volumes of 50 to 125 pages, the series covers a range of content from professional to academic. Typical topics might include:

- A timely report of state-of-the art analytical techniques
- A bridge between new research results, as published in journal article contextual literature review
- A snapshot of a hot or emerging topic
- An in-depth case study or clinical example
- A presentation of core concepts that students must understand in order to independent contributions

SpringerBriefs in Economics showcase emerging theory, empirical research, and practical application in microeconomics, macroeconomics, economic policy, public finance, econometrics, regional science, and related fields, from a global author community.

Briefs are characterized by fast, global electronic dissemination, standard publishing contracts, standardized manuscript preparation and formatting guidelines, and expedited production schedules.

Sohel Rana · Lily Kiminami · Shinichi Furuzawa

Entrepreneurship and Social Innovation for Sustainability

Focusing on a Haor Region of Bangladesh

 Springer

Sohel Rana
Department of Agricultural Economics
and Social Science
Chattogram Veterinary and Animal
Sciences University
Chattogram, Bangladesh

Lily Kiminami
Institute of Science and Technology
Niigata University
Niigata, Japan

Shinichi Furuzawa
Institute of Science and Technology
Niigata University
Niigata, Japan

ISSN 2191-5504   ISSN 2191-5512 (electronic)
SpringerBriefs in Economics
ISBN 978-981-19-7114-3   ISBN 978-981-19-7115-0 (eBook)
https://doi.org/10.1007/978-981-19-7115-0

This Springer imprint is published by the registered company Springer Nature Singapore Pte Ltd.
The registered company address is: 152 Beach Road, #21-01/04 Gateway East, Singapore 189721, Singapore

Acknowledgements

The content of this book is based on the doctoral dissertation entitled "Sustainable Regional Development through Entrepreneurship and Social Innovation: Empirical Analysis on a *Haor* Region of Bangladesh" (Sohel Rana, 2022). This study was conducted with the financial support of the Ministry of Education, Culture, Sports, Science, and Technology of Japan (MEXT Scholarship), and Niigata University, Japan.

Contents

Abbreviations

AEO	Agricultural Extension Officer
AGE	Age
AGES	Age Square
AHI	Annual household income
AMT	Attitudes toward Modern agricultural Technology
BBS	Bangladesh Bureau of Statistics
BCIC	Bangladesh Chemical Industries Corporation
BDT	Bangladeshi Taka
BFP	Bifurcation point
BRAC	Building Resources Across Communities
BS	Block Supervisor
CARE	Cooperative for Assistance and Relief Everywhere
CBDRM	Community-based Disaster Risk Management
CFI	Comparative Fit Index
CIG	Common Interest Group
DAE	Department of Agricultural Extension
DRG	Disaster Risk Governance
DRM	Disaster Risk Management
EDU	Education
EDUS	Education Square
EFP	Equifinality point
EMC	Extension Media Contact
EU	European Union
FGD	Focus group discussion
FS	Farm size
GEM	Global Entrepreneurship Monitor
HSC	Higher Secondary Certificate
IFAD	International Fund for Agricultural Development
INAFI	International Network of Alternative Financial Institutions
KII	Key informants' interview
LGED	Local Government Engineering Department

MITPD	Mitigation and preparedness
NATP	National Agricultural Technology Project
NGO	Non-governmental organization
PCA	Principal component analysis
POPI	People's Oriented Program Implementation
PREVN	Prevention
RECOV	Recovery
REDI	Regional Entrepreneurship and Development Index
RESPO	Response
RMSEA	Root mean square error approximation
SAAO	Sub-Assistant Agriculture Officer
SC	Social capital
SD	Social direction
SEM	Structural equation modeling
SEX	Sex
SG	Social guidance
SI	Social innovation
SSC	Secondary School Certificate
TEM	Trajectory equifinality model
TPE	Total persons engagement
TRAIN	Training
UAO	Upazila Agriculture Officer
UFO	Upazila Fisheries Officer
ULO	Upazila Livestock Officer
VS	Veterinary Surgeon

List of Figures

List of Tables

Chapter 1
Introduction

Abstract Bangladesh's economic development has gradually improved over the last two decades, but there are regional disparities in its development. The Haor region is located in the northeastern part of the country, where the natural environment is unfavorable and socio-economic development is delayed. However, sustainable regional development is a dynamic process that involves multi-sectoral approach to the economic and social development of a region. Each region has its own unique characteristics of natural and human resources, level of technological development, institutional structure, values and ethics. Therefore, the main purpose of this research is to assess the role of entrepreneurship and social innovation for social and cultural changes; and farmers' performance in disaster risk management at the community level towards sustainable regional development in a less favored *haor* region of Bangladesh. The study targets a rural *haor* area of Kishoreganj district of Bangladesh.

Keywords Socio-cultural changes · Entrepreneurship · Social innovation · Disaster risk management · *Haor* region · Bangladesh

1.1 Backgrounds and Main Issues of the Study

Entrepreneurial development is widely recognized as a tool for the economic development of an economy. Drucker (1993) introduced the concept of an entrepreneurial society where innovation and entrepreneurship are normal, steady, and uninterrupted. Moreover, individuals in the entrepreneurial society face a lot of challenges that require substantial social innovation to overcome (Zutshi et al. 2021). Since the early 2000s, social innovation (SI) has been adopted as a steering principle or a reference in both national and international policy documents and policies (Jenson and Harrison 2013; Sabato et al. 2015). It has figured prominently around the world in diverse policy programs to fight poverty, overcome social exclusion and empower minorities (Moulaert and MacCallum 2019). As suggested by Bouchard (2013) that there are at least two major approaches to social innovation: one is concerned with entrepreneurial solutions to address social problems, the other with the collective processes that lead to social change.

© The Author(s), under exclusive license to Springer Nature Singapore Pte Ltd. 2022   1
S. Rana et al., *Entrepreneurship and Social Innovation for Sustainability*,
SpringerBriefs in Economics, https://doi.org/10.1007/978-981-19-7115-0_1

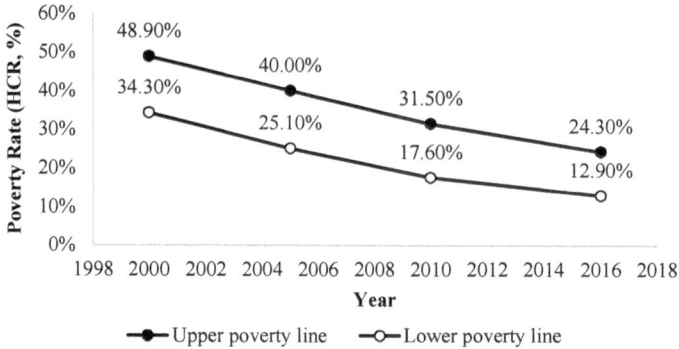

Fig. 1.1 Poverty reduction scenario of Bangladesh (national average). *Data source* BBS (2016)

Bangladesh has achieved continuous economic growth in the last two decades but its advantages are highly uneven in the different regions of the country. There is a lack of regional balance in the resource allocation and policy planning for the development of Bangladesh (Hasnath 2020; Mithun 2021). For example, the country has significantly reduced poverty at the national level (Fig. 1.1) but there is a big regional imbalance among the districts (Table 1.1).

The annual distribution of public budget is influenced by political decisions and allocated according to the priorities of different economic and social sectors rather than the actual needs of the regions especially poor regions. As a result, this unequal distribution of public resources makes a significant variation in economic and social service sectors in different regions of the country. Thus, the regional disparities in the country are getting higher although it has been discussed in the 7th five-year plan (2016–2020) (MoP 2015). Therefore, Bangladesh needs to include systematic regional dimensions with the national sectoral approach in its developmental planning process.

The *haor* region is one of the less favored regions of Bangladesh located in the north-eastern part of the country. The inhabitants in the *haor* region are mostly poor and disadvantaged in terms of getting access to basic livelihood opportunities (MoWR 2012). Livelihood opportunities for the residents of the *haor* region are limited and highly seasonal, as they are mainly concentrated in agriculture (Gillingham 2016). Although agriculture is the main source of livelihood for most of the people in the region, the farmers' entrepreneurship is still limited (Uddin et al. 2019). Moreover, the area faces widespread problems of food insecurity due to disaster risk, limited access to markets and unequal access to resources, etc. (Gillingham 2016). Climate change has severely damaged human and economic developments in Bangladesh's haor regions (Jakariya and Islam 2017).

Additionally, gender discrimination is a root-dependent problem in the *haor* region of Bangladesh (Rana et al. 2020). As a result, the participation of women in local administration (i.e. union council) is particularly marginalized in decision-making at the community level due to the lack of representation and non-cooperation

Table 1.1 Poverty rate by district of Bangladesh

Sl. no	District name (alphabetical order)	Poverty rate (%)	
		Lower poverty line	Upper poverty line
1	Bagerhat	14.4	31.0
2	Bandarban	50.3	63.2
3	Barguna	12.1	25.7
4	Barishal	13.6	27.4
5	Bhola	8.5	15.5
6	Bogra	13.5	27.2
7	Brahmanbaria	4.6	10.3
8	Chandpur	15.3	29.3
9	Chattogram	3.5	13.7
10	Chuadanga	12.1	31.9
11	Cumilla	5.4	13.5
12	Cox's Bazar	7.7	16.6
13	Dhaka	1.7	10.0
14	Dinajpur	45.0	64.3
15	Faridpur	3.2	7.7
16	Feni	3.4	8.1
17	Gaibandha	28.9	46.7
18	Gazipur	1.9	6.9
19	Gopalganj	15.5	29.5
20	Habiganj	9.9	13.4
21	Joypurhat	9.6	21.4
22	Jamalpur	35.2	52.5
23	Jashore	9.0	26.9
24	Jhalokati	9.8	21.5
25	Jhenaidah	12.7	26.5
26	Khagrachari	32.8	52.7
27	Khulna	13.8	30.8
28*	Kishoreganj	34.1	53.5
29	Kurigram	53.9	70.8
30	Kushtia	7.1	17.5
31	Lakshmipur	20.5	32.5
32	Lalmonirhat	23.0	42.0
33	Madaripur	0.9	3.7
34	Magura	37.7	56.7
35	Manikganj	16.3	30.7

(continued)

Table 1.1 (continued)

Sl. no	District name (alphabetical order)	Poverty rate (%)	
		Lower poverty line	Upper poverty line
36	Meherpur	12.4	31.5
37	Maulvibazar	7.0	11.0
38	Munshiganj	1.2	3.1
39	Mymensingh	9.6	22.0
40	Naogaon	18.2	32.2
41	Narail	5.8	16.8
42	Narayanganj	0.0	2.6
43	Narsingdi	4.7	10.5
44	Natore	12.6	24.0
45	Chapai Nababganj	23.7	39.6
46	Netrokona	15.6	34.0
47	Nilphamari	14.2	32.3
48	Noakhali	13.4	23.3
49	Pabna	16.8	33.0
50	Panchagarh	14.2	26.3
51	Patuakhali	24.4	37.2
52	Pirojpur	17.6	32.2
53	Rajshahi	7.3	20.1
54	Rajbari	16.0	33.8
55	Rangamati	10.7	28.5
56	Rangpur	27.0	43.8
57	Shariatpur	5.0	15.7
58	Satkhira	9.3	18.6
59	Sirajganj	12.4	30.5
60	Sherpur	24.3	41.3
61	Sunamganj	19.3	26.0
62	Sylhet	8.8	13.0
63	Tangail	8.6	19.0
64	Thakurgaon	15.5	23.4
Bangladesh average		**12.9**	**24.3**

*Study area

Data source BBS (2016), Household income and expenditure survey 2016, pp. 135–136

in the institutional structures (Gillingham 2016; Kabir et al. 2018). According to the Bangladesh Bureau of Statistics (BBS 2015), women's engagement in economic activities is lower in the *haor* area (14.85%) than the national average (16.54%). Social tolerance is a barrier for rural women to involve in economic activities in the haor region of Bangladesh (Rana et al. 2021) although initiatives are taken by various stakeholders. In addition, achieving gender equality and women's empowerment is a challenge for Goal 5 (Gender equality) of the SDGs based on international consensus. Although gradual improvement has been seen in Bangladesh in recent years, significant challenges remain (Sachs et al. 2021).

Moreover, Cooperative for Assistance and Relief Everywhere (CARE) Bangladesh has executed several programs in the last decade in the *haor* region of Bangladesh to support the *haor* people for improving their livelihood and food security and strengthening household ability to respond to development opportunities. Based on the findings, CARE Bangladesh has identified a series of programs to be implemented as priorities for advancing the poor community of the *haor* region through women's empowerment, inclusive governance, livelihood strengthening, building resilience to disasters and promoting climate change adaptation, sanitation and hygiene; and maternal and child nutrition and health care from 2015 to 2020 (Gillingham 2016). In addition, the master plan of the government for the *haor* region has been formulated to achieve the following goals such as economic development, food security, decent standard of living for the people, poverty alleviation, and protection of the natural environment (MoWR 2012). However, as pointed out by BRAC, injustice, and discrimination are the most important reasons for perpetuating poverty in the *haor* region.

In addition, the *haor* region is one of the most disaster-prone regions of the country. The region becomes waterlogged from May to October every year and is at risk of flooding during the pre-monsoon season, with a severe negative impact on the boro rice harvest (April) which forms the basis of livelihood for most of the people in the region. Therefore, disaster risk management is a very important issue for the regional development of *haor* region of Bangladesh while it may not so important for other regions of the country. Although the government of Bangladesh had set the vision for disaster risk management, and the country has made significant development gains over the last decade that reduced the mortality resulting from regular disasters (Fig. 1.2). However, the traditional disaster counter-measures in Bangladesh were based on the concept of providing post-disaster relief and rehabilitation facilities that have failed to make any contribution toward the development process (EC 2007). In addition, the counter disaster policies of Bangladesh emphasized more on the top-level linking network e.g. national and international organizations, foreign states, and donors, etc. (Islam and Walkerden 2017) than fully addressing the local needs, capacities, and socioeconomic conditions (Asgary and Halim 2011), and mentioned no links with other stakeholders at the grassroots level (Islam and Walkerden 2017). Therefore, social innovation is considered as an effective solution to solve social problems (Phills et al. 2008) and empower marginalized people (Moulaert al. 2013).

The main purpose of this study is to analyze the relationship between social capital, empowerment and technology adoption in the entrepreneurial ecosystem;

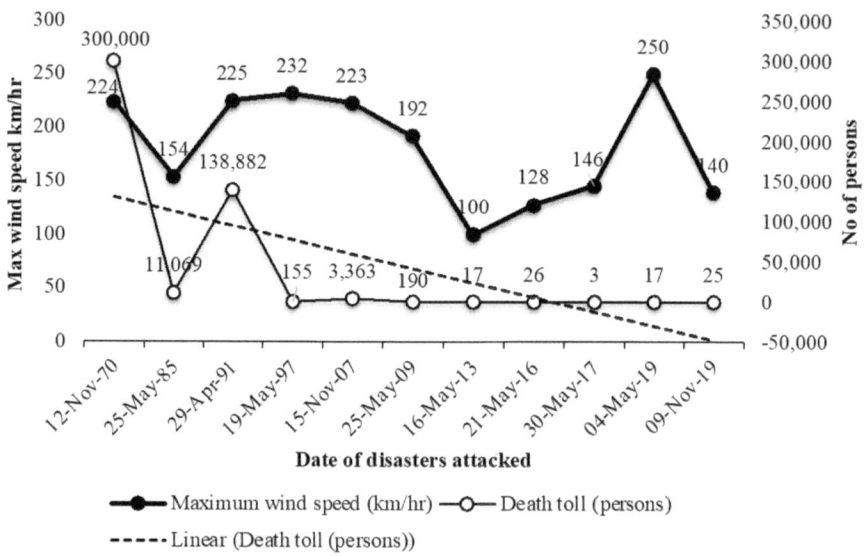

Fig. 1.2 Major cyclones hit the Bangladesh (*Source* Bangladesh metrological department)

and farmers' performance in disaster risk management at community level towards sustainable regional development in a less favored *haor* region of Bangladesh. It will also draw policy implications based on the analytical results for the local and central governments on their policy and practice for enhancing sustainable regional development of the *haor* region with the mainstream development of the country. To reach the purpose of this research, both qualitative (Trajectory Equifinality Modeling) and quantitative (Structural Equation Modeling) analytical approaches will be applied based on primary data.

1.2   Layout of the Book

The layout of the book is as shown in Fig. 1.3. First, a literature review on the theoretical approaches to regional development, the regional disparity in Bangladesh, entrepreneurship and social innovation in sustainable regional development, disaster risk management at the community level, and related areas will be undertaken. Secondly, we will describe the regional characteristics of the target area compared with the national socio-economic situation of the country to understand the contextual situation of the study region. Thirdly, the hypotheses for this research will be set based on the literature review and the existing situation of the issues discussed above in the study area perspective, and potential methodologies will be introduced for hypotheses verification. Fourthly, empirical analysis for hypotheses verification will be undertaken by integrating both quantitative and qualitative analytical approaches.

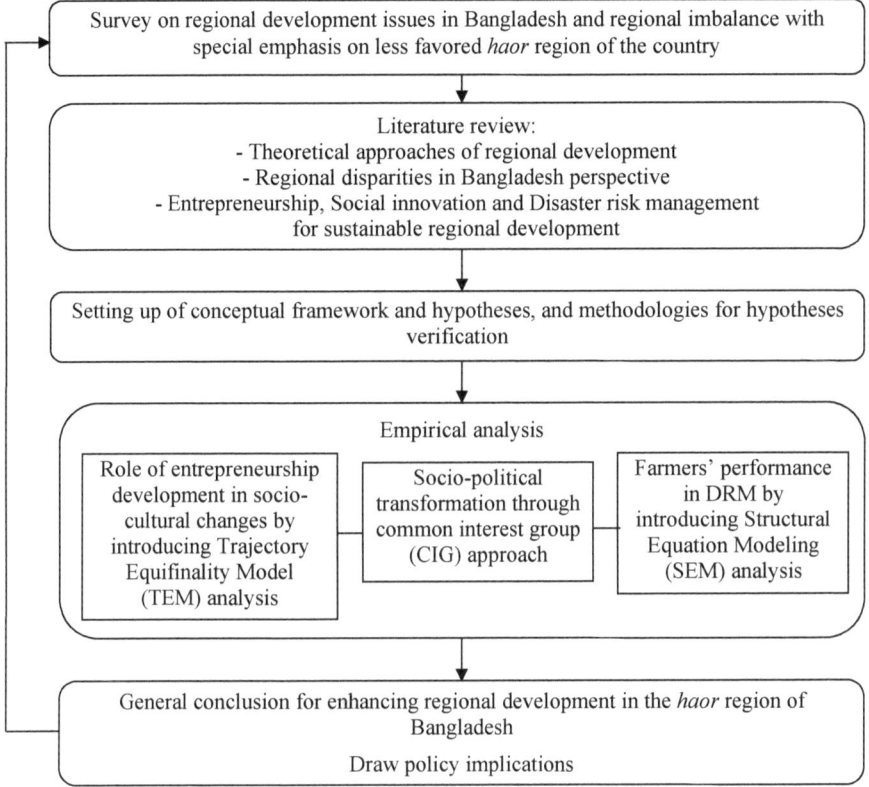

Fig. 1.3 Layout of the book

Finally, a general conclusion and policy implications will be drawn for enhancing sustainable regional development in the *haor* region of Bangladesh based on the analytical results.

References

Asgary A, Halim MA (2011) Measuring people's preferences for cyclone vulnerability reduction measures in Bangladesh. Disaster Prev Manag 20(2):186–198

BBS (2015) Economic census 2013. Bangladesh Bureau of Statistics, Ministry of Planning. Government of the People's Republic of Bangladesh, Dhaka

BBS (2016) Report on the household income and expenditure survey 2016. Bangladesh Bureau of Statistics, Ministry of Planning, Government of the People's Republic of Bangladesh, Dhaka

Bouchard M (2013) Innovation and the social economy: The quebec Experience. University of Toronto Press, Toronto

Drucker PF (1993) Innovation and entrepreneurship practice and principles. Harper Business, New York

EC (2007) Vulnerabilities and capacities of people to cope with disaster. European Commission. Handicap International, Bangladesh

Gillingham S (2016) CARE Bangladesh program strategy: haor region 2015–2020. CARE Bangladesh, Dhaka

Hasnath SA (2020) Uneven development in Bangladesh: a temporal and regional analysis. In: Chen Z, Bowen W, Whittington D (eds) Development studies in regional science. New frontiers in regional science: Asian perspectives, vol 42. Springer, Singapore, Pte Ltd, pp 199–219

Islam R, Walkerden G (2017) Social networks and challenges in government disaster policies: A case study from Bangladesh. International Journal of Disaster Risk Reduction 22:325–334

Jakariya M, Islam MN (2017) Evaluation of climate change induced vulnerability and adaptation strategies at Haor areas in Bangladesh by integrating GIS and DIVA model. Model Earth Syst Environ. https://doi.org/10.1007/s40808-017-0378-9

Jenson J, Harrisson D (2013) Social innovation research in the European Union: approaches, findings and future directions. Policy review, directorate general for research and innovation. Socio-economic sciences and humanities. Luxembourg: Publications Office of the European Union

Kabir SMS, Aziz MA, Shathi AKMSJ (2018) Women empowerment and governance in Bangladesh. Indian Journal of Women and Social Change 3(1):24–35

Mithun MMZ (2021) Regional development planning and disparity in Bangladesh. E3 Journal of Business Management and Economics 11(1):10–020

MoP (2015) 7th five-year plan FY2016-FY2020: accelerating growth; empowering citizens. Ministry of Planning, Government of the People's Republic of Bangladesh, Dhaka

Moulaert F, MacCallum D (2019) Advanced introduction to social innovation. Edward Elgar Publishing, Cheltenham, UK and Northampton, MA, USA

Moulaert F, MacCallum D, Mehmood A, Hamdouch A (2013) (eds), The international handbook on social innovation: collective action, social learning and trans-disciplinary research. Edward Elgar Publishing, Cheltenham, UK and Northampton, MA, USA

MoWR (2012) Master plan of haor area. Bangladesh Haor and Wetland Development Board, Ministry of Water Resources, Government of the People's Republic of Bangladesh, Dhaka

Phills JA, Deiglmeier K, Miller DT (2008) Rediscovering social innovation. Stanf Soc Innov Rev 6(4):34–43

Rana S, Kiminami L, Furuzawa S (2020) Analysis on the factors affecting farmers' performance in disaster risk management at community level: focusing on a haor locality in Bangladesh. Asia-Pacific Journal of Regional Science 4(3):737–757

Rana S, Kiminami L, Furuzawa S (2021) Social innovation for women's empowerment in disaster risk governance: focusing on common interest groups in the *haor* region of Bangladesh. Studies in Regional Science 51(1):145–155

Sabato S, Vanhercke B, Verschraegen G (2015) The EU framework for social innovation between entrepreneurship and policy experimentation. ImPRovE Working Paper 15/21

Sachs J, Traub-Schmidt G, Kroll C, Lafortune G, Fuller G (2021) Sustainable development report 2021: The decade of action for the sustainable development goals includes the SDG index and dashboards. Cambridge University Press, Cambridge

Uddin MT, Hossain N, Dhar AR (2019) Business prospects and challenges in haor areas of Bangladesh. Journal of Bangladesh Agricultural University 17(1):65–72

Zutshi A, Mendy J, Sharma GD, Thomas A, Sarker T (2021) From challenges to creativity: enhancing SMEs' resilience in the context of COVID-19. Sustainability 13:6542. https://doi.org/10.3390/su13126542

Chapter 2
Literature Review on Regional Development, Entrepreneurship, Social Innovation and Disaster Risk Management

Abstract Based on the literature review on regional development, entrepreneurship, social innovation and disaster risk managment, this study pointed out that although the accumulation of studies on entrepreneurship and social entrepreneurship are advanced, the study focusing on entrepreneurship and socio-cultural change in a specific cultural context in developing countries by comparing male and female entrepreneurs is scarce. In addition, there are quite a few research on the impacts of common interest group (CIG) approach for empowering women through socio-political transformation in the *haor* region. Therefore, comparing the cases of local male and female entrepreneurs from various angles in the *haor* region of Bangladesh will help us to have a deeper understanding of the above-mentioned issues and to consider effective approaches for the solution. Furthermore, it is necessary to identify multi-causal relationships among factors affecting farmers' performance in disaster risk management (DRM) at community level through the covariance structure analysis.

Keywords Regional development · Entrepreneurship · Social innovation · Common interest group · Disaster risk management

2.1 Theoritical Approaches of Regional Development

Regional development is a dynamic process that involves a multi-sectoral approach to the economic and social development of a region. Each region has its own uniqueness and unique characteristics of natural and human resources, level of technological development, capital, knowledge, institutional and legislative structure, values and ethics (Sabic and Vujadinovic 2017). According to the theory of the Regional System of the Man-Land Relationship, the combination and rational use of different resources is essential to produce regional differences that promote sustainable development in a region (Wang 2017). Kiminami and Kiminami (2017) explained the relationship between regional development and sustainability (Fig. 2.1) in terms of total economic, social and environmental value created at a certain period of time in a region through various activities. If the minimum sutainablity in a region is set at (A_{min}, B_{min}) and the frontiers area $(A_{t0}B_{t0})$ is assumed in the region created at time t0 as the total

economic, social and envornmental value. The innovations and entrepreneurship in
the region are supposed to be created at the time of t1, as the main drivers to enlarge
the frontiers area to $(A_{t1} B_{t1})$, as a result the level of sustainability would be shifted
to $P_{t1}*$ from $P_{t0}*$ which denotes that there has been sustainable development in the
region. In addition, the rural and agriculture development in a region will not be
achieved without paying attention to the development of agriculture as an industry.

In that case the developments of entrepreneurship play a vital role to transform the
traditional agro-farming system as an industry at the local level to diversify the rural
economy. EU (2014) pointed out that the contextual factors regulate the systems of
entrepreneurship in a particular region, although entrepreneurial activities are undertaken individually and they are embedded in a specific regional context. However,
regional system needs to be discussed under the consideration of changing the existed
power relationship between man and women especially in rural communities. As the
regional development prolems are complex, context specific, and multidimensional
in nature; the purpose of regional planning is to reduce poverty, income inequalities
and regional imbalances in the development process.

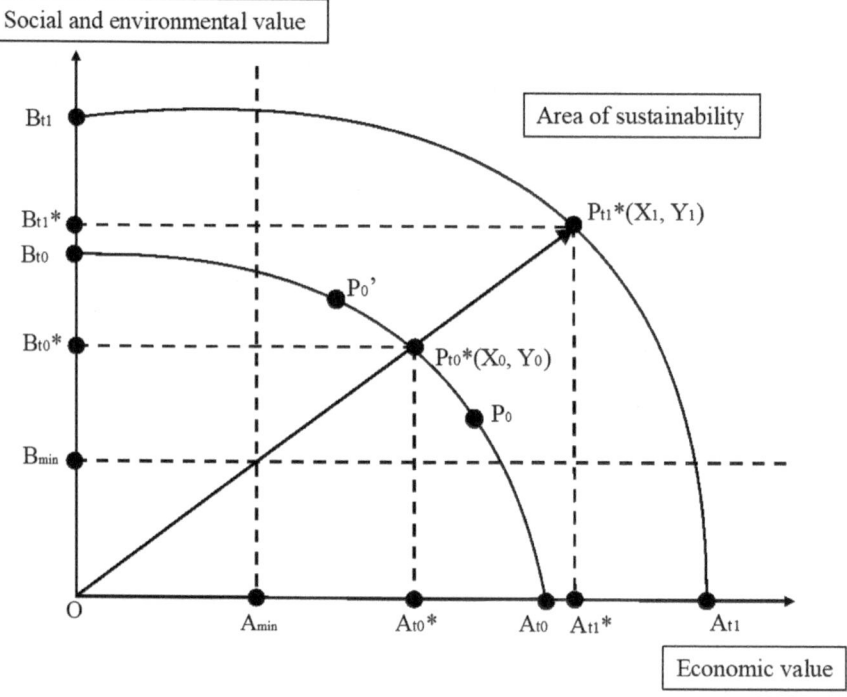

Fig. 2.1 Relationship between regional development and sustainability (*Source* Based on Kiminami and Kiminami (2017) Fig. 1, p. 549)

2.2 Regional Disparities in Bangladesh Perspective

In 2018, Bangladesh has been graduated from least developed country (LDC) to the status of developing country,[1] but regional balance in the development of Bangladesh has not received due attention in the development process (Haque 2005). Although *haor* region is one of the most disaster-prone regions of the Bangladesh but disaster recovery support is less in the region than in the national average (BBS 2015). Haque (2005) conducted a spatial analysis on the regional development of Bangladesh through the input–output model and concluded that the economic sectors interact more strongly in the developed regions than in the least developed regions of the country and the impact of investment in major economic sectors is less in under developed regions than in developed regions. According to Hasnath (2020) there are income inequality of the households and the disparities among the different regions of Bangladesh. Mithun (2021) conducted a spatial analysis on regional disparities in Bangladesh and concluded that some regions of Bangladesh are still lagging far behind the mainstream national development because of the fact that they continuously receive marginal share of public expenditure and country's national budget.

2.3 Enrepreneurship in Poverty and Empowerment Dynamics

The Global Entrepreneurship Monitor (GEM) defines entrepreneurship as any effort to start a new business or to expand an existing business by an individual or a group which create opportunities for the society (Reynolds et al. 1999). Thus, entrepreneurship means not just starting a business but also a source of economic and social empowerment. Because entrepreneurs both male and female have to face different kinds of challenges and entrepreneurs have to decide to be self-independent in the process of entrepreneurial ecosystem and contribute to the society such as create values, create jobs, poverty reduction etc. Many studies have strongly argued that there is a strong positive relationship between entrepreneurial development and poverty reduction (Naminse and Zhuang 2018; Si et al. 2015; Bruton et al. 2015; Morris et al. 2020; Polak 2009).

Women's empowerment process enhances their control to strategic life choices through increased access to economic, political and socio-cultural domains in the society (Chen and Tanaka 2014). Entrepreneurial activities not only empower women, but also contribute to changing social structure in the communities towards sustainable regional development. Facilitating women's economic activity (Haugh and Talwar 2016) through cognitive change in dimensions of empowerment and

---

[1] https://cpd.org.bd/bangladesh-transitioning-to-developing-country/ (Accessed on 28 April 2021).

entrepreneurship (Santos et al. 2019; Santos and Neumeyer 2021) causes dynamic changes in the nature of the society (Johnson and Schaltegger 2020).

Social entrepreneurship is considered as a means of addressing social and environmental problems towards sustainable development (Bansal et al. 2019). Social entrepreneurship is defined as a business initiative that is primarily concerned with recognizing social problems and effort to make positive changes in society through entrepreneurial activities (Austin et al. 2006). Furthermore, social entrepreneurs can create innovations even in resource constrained environment (Kickul et al. 2018). Women entrepreneurs are helping to change the subsequent rebellion against gender relationships and the traditional culture of subjugation in male dominate society. In this regard, society needs a socio-cultural transformation through changing collective cognition (Moulaert and MacCallum 2019).

2.4 Social Capital and Role of Technology Adoption in Entrepreneurship

In our study we have considered the definition of social capital as "the features of social organization, such as trust, norms, and networks that can improve the efficiency of the society by facilitating coordinated actions" (Putnam 1993, p. 167). Neumeyer et al. (2019a) clarified that the level of social capital of male and female entrepreneurs differs significantly depending on the type of enterprise, race, and ethnicity. In addition, entrepreneurial network connections with different stakeholders also differ across different social clusters that significantly affect their entrepreneurial development (Neumeyer et al. 2019b). Ferdousi and Mahmud (2019) noted that social business has an important role to play in the development of women entrepreneurs in rural Bangladesh through the promotion of financial, human, and social capital.

However, there is still a lack of specific legal policy infrastructure for social enterprises in Bangladesh, although other sectoral policies affect the activities of social enterprises (Darko 2016). Darko (2016) recommended that three key areas are important for the growth of social enterprises in Bangladesh such as access to finance, entrepreneurial skills development, and raising awareness among various stakeholders. In addition, technology adoption can play an important role and increase the efficiency of entrepreneurial activities of small enterprises (Rosin et al. 2020) in long-term entrepreneurial sustainability (Neumeyer et al. 2021; Orser et al. 2019). However, the adoption of information technologies (IT) among SMEs (small and medium enterprises) is largely influenced by their gender (Orser and Riding 2018).

2.5 Common Interest Group (CIG) Approach and Social Innovation

The common interest group (CIG) is an association of rural farmers with similar socio-economic backgrounds who share a common interest or passion (e.g. cultivation of the same crop) and exchange thoughts, ideas and beliefs within a particular community (INAFI 2016; Rana et al. 2018). The formation of CIG was started in Bangladesh in 2009–2010 under the National Agricultural Technology Project (NATP). NATP is a multi-agency project in Bangladesh jointly funded by the World Bank, IFAD, USAID, and the Government of Bangladesh. The overall goal of NATP is to support Bangladesh's strategies to improve national agricultural productivity and farm income by revitalizing the national agricultural technology system with an emphasis on small and marginal farmers. The first phase of NATP was started in 2008 and completed in 2014 which formed 20,012 CIGs covering 120 upazillas (sub-districts) under 25 districts of Bangladesh. Following the evaluation of the first phase, the second phase of NATP was started in October 2016 and will end in September 2021 which has constituted 40,710 CIGs covering 270 upazilas of 57 districts of the country.

The study area was included in the 2nd phase of NATP.[2] The local agriculture offices formed the CIGs in the specific areas with the help of village leaders. Each CIG consists of 20–30 members. Although there are some specific criteria for selecting farmers as members of the CIG those who are interested in working in a group and especially small and marginal farmers (1–3 acre farm size) were given preference. Ten CIGs were formed (6 male CIGs and 4 female CIGs) in a union (the smallest unit of local government and rural administration). In the study area, CIGs were formed in 2017. Each CIG has an executive committee of nine members which is responsible for the overall management of CIG. They sit in a monthly meeting and discuss contemporary farming issues in the locality and relevant agricultural technologies in the presence of SAAO (Sub-Assitant Agriculture Officer). They received training on improved farming practices and take part in technological demonstrations organized by the local agriculture office. They maintain a bank account and deposit a small amount of money every month as saving (e.g. 100 Taka per member per month, it varies on the financial conditions of the group members) to build a self-contributory fund. The CIG farmers' group also received subsidy (e.g. to buy farm machinery) from the Ministry of Agriculture. The CIGs usually invest their savings in various production and post-production activities and small business purposes. The CIG approach has created higher opportunities for rural farmers in Bangladesh to significantly improve their livelihoods (Rana et al. 2018; IFAD 2016; INAFI 2016). However, the CIG approach has some limitations as it involves a small segment (about 10%) of farmers in the community. Furthermore, CIGs may lose their main purpose of the association if they are not managed properly by the concerned local agriculture offices.

---

[2] Source: NATP website (https://natp2pmu.portal.gov.bd/site/page/4c2e8dee-9916-4eb8-99b3-60e 1886e1e57/.) (Accessed on 12 October 2020).

Social innovation is defined as a new process or product or service that meets social needs through the integration of organizations and transforms the socio-political structure that leads to the collective empowerment of communities (Moulaert and MacCallum 2019). In addition, social innovation is understood as both a process and outcome of social institutional change based on the institutional theory framework (Logue 2020). In this sense, if the CIG approach plays a role in enabling rural women entrepreneurs to increase their farm productivity and income, to access new knowledge and technology, local institutions and leadership skills at the organizational level etc., it can be recognized as an effective approach to creating social innovation in the region.

2.6 Disaster Risk Management (DRM) at Community Level

In the event of a disaster, local communities and governments are often the first responders before the support from outsiders. Therefore, more attention should be paid for the enhancement of DRM practices at the local level (Delica-Willison 2005; Kafle and Murshed 2006). A community-based approach in disaster risk management (CBDRM) was strongly recommended by Maskrey (1989) and Shaw (2012) focused on the activities related to disaster risk reduction undertaken by local communities on a broader perspective. CBDRM is a participatory process aimed at the identification, assessment, treatment and planning for hazards and vulnerabilities of various kinds through active participation of communities (Krummacher 2014). CBDRM hence signify the active and continued involvement of communities in the decision-making process of disaster risk reduction (Niekerk et al. 2018). UNDP (2016) proposed the elements for CBDRM implementation are (i) the existence of a local disaster risk management committee or organization (ii) community hazard, vulnerability and capacity/resources mapping (iii) a community disaster risk management plan (iv) training in disaster risk management and community-learning systems (v) regular community simulation and exercises (vi) early warning system and (vii) a disaster risk reduction fund. Niekerk et al. (2018) concluded that CBDRM has been widely practiced across the world although its implementation varies in regions. In Asia, information dissemination and capacity building of communities has been a common element of CBDRM where NGOs play a significant role to CBDRM activities through their work in facilitating capacity building and development of livelihood related skills.

Poverty: The poor households in Bangladesh are extremely vulnerable due to the loss of income and employment caused by natural disasters (Akter and Mallick 2013). Therefore, a sustainable livelihood approach that integrated poverty reduction and disaster risk management to achieve livelihood outcomes with more income, increased well-being, improved food security and reduced vulnerability of disasters is mentioned (Cannon et al. 2003; Schmidt et al. 2005).

Social capital: Social capital, such as networks, norms, and trust, that facilitate coordination and cooperation for mutual benefit (Putnam 1993) has positive roles in different stages of disaster risk management (Chui et al. 2014; Islam and Walkerden 2014), significantly affects the successful implementation of disaster-related policy (Joshi and Aoki 2014). However, Islam and Walkerden (2015, 2017) documented that the existing governmental policies of DRM in Bangladesh pay no attention to social capital during the distribution of relief goods in particular. As a result, relief dependency at the local level is increasing.

Gender: Women are considered as necessary stakeholders and have important contribution for managing disasters at different level (Nasreen 2012; Cvetkovic et al. 2018). The absence of women in disaster management planning creates many issues during response and recovery stages because they are considered as agents of change within their communities and even in wider spectrum (Ikeda 2009; Morchain and Kelsey 2016). However, the low mobility of women controlled by different socio-cultural issues make them have low access to productive resources and income opportunities, which lead them to have a week decision making power in Bangladesh (Mehta 2007). In spite of the advancement in disaster reduction, women-sensitive policy for DRM is still lacking in Bangladesh (Ahmed 2019).

2.7 Summery of the Literature Review

The originality of this study is as follows. Although the accumulation of literature on entrepreneurship and social entrepreneurship research are advanced, the study focusing on entrepreneurship and socio-cultural change in a specific cultural context in developing countries by comparing male and female entrepreneurs is scarce. First, both male and female entrepreneurs have high motivation, but their performances seem to be different. Therefore, comparing the cases of local male and female entrepreneurs from various angles in the *haor* region of Bangladesh will help us to have a deeper understanding of the above-mentioned issues and to consider effective approaches for the solution. Secondly, although the common interest group (CIG) is a widely practiced approach in the rural areas of Bangladesh, there are quite a few research on the impacts of CIG approach for empowering women through socio-political transformation in the *haor* region. Thirdly, departing from the most of the existing literatures, in which factors affecting performance in DRM are identified through multiple regression analysis (simple-line causal relations between factors and performance in DRM), it is necessary to identify multi causal relationships among factors affecting farmers' performance in DRM at community level through the covariance structure analysis, etc. in order to obtain fruitful policy implications for DRM, since the factors are dependent on each other in a complex and specific manner in a certain community like *haor* areas of Bangladesh.

References

Ahmed S (2019) A gender sensitive policy framework for disaster management in Bangladesh. A PhD thesis, Institute for Sustainable Industries and Livable Cities, Victoria University, Melbourne, Australia

Akter S, Mallick B (2013) The poverty–vulnerability–resilience nexus: evidence from Bangladesh. Ecol Econ 96:114–124

Austin J, Stevenson H, Wei-Skillern J (2006) Social and commercial entrepreneurship: same, different, or both? Entrep Theory Pract 30(1):1–22

Bansal S, Garg I, Sharma GD (2019) Social entrepreneurship as a path for social change and driver of sustainable development: a systematic review and research agenda. Sustainability 11:1091. https://doi.org/10.3390/su11041091

BBS (2015) Bangladesh disaster-related statistics (2015) Climate change and natural disaster perspectives. Bangladesh Bureau of Statistics, Ministry of Planning, Government of the People's Republic of Bangladesh, Dhaka

Bruton GD, Ahlstrom D, Si S (2015) Entrepreneurship, poverty, and Asia: moving beyond subsistence entrepreneurship. Asia Pacific J Manag 32(1):1–22

Cannon T, Twigg J, Rowell J (2003) Social vulnerability, sustainable livelihoods and disasters. London: Department for International Development (DIFD)

Chen YZ, Tanaka H (2014) Women's empowerment. In Michalos AC (ed) Encyclopedia of quality of life and well-being research. Springer, Dordrecht

Chui C, Feng JY, Jordan L (2014) From good practice to policy formation- the impact of third sector on disaster management in Taiwan. Int J Disaster Risk Reduct 10:28–37

Cvetkovic VM, Roder G, Ocal A, Tarolli P, Dragicevic S (2018) The role of gender in preparedness and response behaviors towards flood risk in Serbia. Int J Environ Res Public Health 15:1–21

Darko E (2016) Social enterprise policy landscape in Bangladesh, ODI report

Delica-Willison Z (2005) Community-based disaster risk management: local level solutions to disaster risks. Tropical Coasts 12(1):66–73

EU (2014) REDI: the regional entrepreneurship and development index-measuring regional entrepreneurship. Final report. Publication office of the European Union, Luxembourg

Ferdousi F, Mahmud P (2019) Role of social business in women entrepreneurship development in Bangladesh: perspectives from nobinudyokta projects of Grameen Telecom Trust. J Glob Entrep Res 9:58

Hasnath SA (2020) Uneven development in Bangladesh: a temporal and regional analysis. In: Chen Z, Bowen W, Whittington D (eds) Development studies in regional science. New frontiers in regional science: Asian perspectives, vol 42. Springer, Singapore, Pte Ltd, pp 199–219

Haugh HM, Talwar A (2016) Linking social entrepreneurship and social change: the mediating role of empowerment. J Bus Ethics 133(4):643–658

Haque A (2005) Regional development in Bangladesh: a spatial analysis through input-output model. A PhD thesis, Department of urban and regional planning, Bangladesh University of Engenering and Technology, Dhaka, Bangladesh

IFAD (2016) Project completion report validation. National agricultural Technology Project, People's Republic of Bangladesh.

Ikeda K (2009) How women's concerns are shaped in community-based disaster risk management in Bangladesh. Contemp South Asia 17(1):65–78

INAFI (2016) Safe vegetable cultivation under National Agricultural Technology Project: Phase-I; The case of crop CIGs in Kuliar Char, Kishoreganj. http://asia.procasur.org/wp-content/uploads/2016/10/NATP_Safe-Vegetables.pdf. Accessed 16 September 2020

Islam R, Walkerden G (2014) How bonding and bridging networks contribute to disaster resilience and recovery on the Bangladeshi coast. Int J Disaster Risk Reduct 10:281–291

Islam R, Walkerden G (2015) How do links between households and NGOs promote disaster resilience and recovery?: a case study of linking social networks on the Bangladeshi coast. Nat Hazards 78(3):1707–1727. https://doi.org/10.1007/s11069-015-1797-4

Islam R, Walkerden G (2017) Social networks and challenges in government disaster policies: a case study from Bangladesh. Int J Disaster Risk Reduct 22:325–334

Johnson M, Schaltegger S (2020) Entrepreneurship for sustainable development: a review and multilevel causal mechanism framework. Entrep Theory Pract 44:1141–1173

Joshi A, Aoki M (2014) The role of social capital and public policy in disaster recovery: a case study of Tamil Nadu State, India. Int J Disaster Risk Reduct 7:100–108

Kafle SK, Murshed Z (2006) Community-based disaster risk management for local authorities: participant's workbook. Pathumthani, Thailand: Asian Disaster Preparedness Center (ADPC)

Kickul J, Griffiths M, Bacq S, Garud N (2018) Catalyzing social innovation: is entrepreneurial bricolage always good? Entrep Reg Dev 30(3&4):407–420

Kiminami L, Kiminami A (2017) Rural and agriculture development in regional science. In Shibusawa H, Sakurai K, Mizunoya T, Uchida S (eds) Socioeconomic environmental policies and evaluations in regional science, New frointiers in regional science: Asian perspectives 24, Springer Science Singapore, pp 537–555

Krummacher A (2014) Community based disaster risk management. Vienna: 22nd OSCE economic and environmental forum

Logue D (2020) Theories of social innovation. Edward Elgar Publishimg, Cheltenham, UK and Northamton, MA, USA

Maskrey A (1989) Disaster mitigation: a community-based approach. Oxfam, Oxford, UK

Mehta M (2007) Gender matters: lessons for disaster risk reduction in South Asia. The international center for integrated mountain development (ICIMOD)

Mithun MMZ (2021) Regional development planning and disparity in Bangladesh. J Bus Manag Econ 11(1):010–020

Morchain D, Kelsey F (2016) Findings ways together to build resilience: the vulnerability and risk assessment methodology. Oxfam UK

Morris MH, Santos SC, Neumeyer X (2020) Entrepreneurship as a solution to poverty in developed economies. Bus Horiz 63(3):377–390

Moulaert F, MacCallum D (2019) Advanced introduction to social innovation. Edward Elgar Publishing, Cheltenham, UK and Northampton, MA, USA

Naminse EY, Zhuang J (2018) Does farmer entrepreneurship alleviate rural poverty in China? evidence from Guangxi Province. Plos ONE 13(3):1–18

Nasreen M (2012) Women and girls vulnerable or resilient? University of Dhaka, Bangladesh, Institute of Disaster Management and Vulnerability Studies

Neumeyer X, Santos SC, Caetano A, Kalbfleisch P (2019a) Entrepreneurship ecosystems and women entrepreneurs: a social capital and network approach. Small Bus Econ 53(2):475–489

Neumeyer X, Santos SC, Morris MH (2019b) Who is left out: exploring social boundaries in entrepreneurial ecosystems. J Technol Transf 44(2):462–484

Neumeyer X, Santos SC, Morris MH (2021) Overcoming barriers to technology adoption when fostering entrepreneurship among the poor: the role of technology and digital literacy. IEEE Trans Eng Manag 68(6):1606–1618

Niekerk DV, Nemakonde LD, Kruger L, Genade KF (2018) Community-based disaster risk management. In Rodriguez et al. (ed) Handbook of disaster research, handbooks of sociology and social research, 2nd edn. Springer, Switzerland, pp 411–429

Orser B, Riding A (2018) The influence of gender on the adoption of technology among SMEs. Int J Entrep Small Bus 33(4):514–531

Orser B, Riding A, Li Y (2019) Technology adoption and gender-inclusive entrepreneurship education and training. Int J Gend Entrep 11(3):273–298

Polak P (2009) Out of poverty: what works when traditional approaches fail. Berrett-Koehler Publishers, Inc., San Francisco

Putnam RD (1993) Making democracy work: civic traditions in modern Italy. Princeton University Press, Princeton

Rana MM, Farouque MG, Rahman MZ (2018) Change of livelihood status of common interest group members: interventions of National Agricultural Technology Program. Bangladesh J Extension Educ 30(2):37–46

Reynolds PD, Hay M, Camp SM (1999) Global entrepreneurship monitor: 1999 Executive report, Kaufman Center for Entrepreneurial Leadership

Rosin AF, Proksch D, Stubner S, Pinkwart A (2020) Digital new ventures: assessing the benefits of digitalization in entrepreneurship. J Small Bus Strateg 30(2):59–71

Sabic D, Vujadinovic S (2017) Regional development and regional policy. Collection of Papers—Faculty of Geography at the University of Belgrade 65(1):463–477

Santos SC, Neumeyer X (2021) Gender, poverty and entrepreneurship: a systematic literature review and future research agenda. J Dev Entrepreneurship 26(3):215008–1–31

Santos SC, Neumeyer X, Morris MH (2019) Entrepreneurship education in a poverty context: an empowerment perspective. J Small Bus Manage 57(1):6–32

Schmidt A, Bloemertz L, Macamo E (2005) Linking poverty reduction and disaster risk management. Eschborn: German Technical Co-operation (GTZ)

Shaw R (2012) Community based disaster risk reduction. Bingley, Emerald

Si S, Yu X, Wu A, Chen S, Chen S, Su Y (2015) Entrepreneurship and poverty reduction: a case study of Yiwu. China. Asia Pacific J Manag 32(1):119–143

UNDP (2016) Myanmar community-based disaster risk management manual. UNDP, Naypyidaw

Wang Z (2017) Principles of regional science. Springer Nature Singapore Pte Ltd

Chapter 3
Regional Characteristics of the Target Region

Abstract The *haor* region of Bangladesh has a wetland ecosystem. Most linking roads become submerged from May to October of the year and movement is mainly possible by boat only. However, motorized vehicle transportation is possible only in major towns of that region. People face greater challenges in transportation than those in other parts of the country. In this chapter, we explain the socio-economic developmental situation of the target region as compared to national development based on the government statistical data. It is clarified that the target area is behind the national average level in almost all the socio-economic development indicators such as literacy rate, household income, gender equality, and institutional collaboration etc. Although, the *haor* region covers about 12% of the total geographical area, it contributes about 22% of the total rice (staple food) production of the country. Poverty is more acute in the *haor* region due to lack of market access and power supply, and limited presence of financial institutions.

Keywords Agricultural production · Socio-economic development · *Haor* region · Kishoreganj district · Bangladesh

3.1 Contribution of *Haor* Region to National Food Security

This study was conducted in a rural area (Itna union) of Kishoreganj district (one of the seven *haor* districts) of Bangladesh (Fig. 3.1). The regional characteristics of the study area are described as follows.

The *haor* region of Bangladesh covers about 12% of the total geographical area of the country (BBS 2011) but contributes 21.44% of the total rice (staple food) production of the country (Table 3.1). Although the *haor* region contributes a lot to national food security, it lags far behind socio-economically compared to mainstream national development. However, the Government of Bangladesh has started providing assistance to farmers in poor *haor* areas by subsidizing 70% of the purchase price of agricultural machinery to promote sustainable agricultural production system in the region (MoA 2020).

© The Author(s), under exclusive license to Springer Nature Singapore Pte Ltd. 2022   19
S. Rana et al., *Entrepreneurship and Social Innovation for Sustainability*,
SpringerBriefs in Economics, https://doi.org/10.1007/978-981-19-7115-0_3

Fig. 3.1 Map of Bangladesh showing *haor* region and target area (*Source* Banglapedia)

Table 3.1 Contribution of *haor* region in country's staple food (rice) production

Year	Country's total rice production (Thousand M. Ton)	Rice production in *haor* region (Thousand M. Ton)	% Of national total
2015–16	34,710	5,209.54	15.01
2016–17	33,804	4,622.96	13.67
2017–18	36,279	5,652.50	15.58
2018–19	28,455	5,737.69	20.16
2019–20	28,213	6,048.24	21.44

Source Authors calculation based on the data from Yearbook of agricultural statistics of Bangladesh 2016, 2017, 2018, 2019, 2020, BBS

3.2 Household Characteristics in the Target Region

In the study area, about 52% of households are farm households who depend on agriculture for their livelihood. The majority of farm families (84.64%) fall into the category of small farm families with farm sizes below 2.50 acres (Table 3.2).

Among the non-farm households, a significant portion (37.50%) are agricultural labor who do not own farm land. They sell their labor to farming families. However, they face a lack of job opportunities during the monsoon season (May to October) when there is no or very limited farming operations in the region. Although fishing is an alternative means of livelihood for them in the rainy season, people are not able to get easy access to open water bodies (Gillingham 2016).

3.3 Land Utilization Pattern in Relation to Agricultural Production

The farming community in the haor region gets about half the opportunity to cultivate their land as compared to other parts of the country due to water logging in the region. From May to October most of the land goes under water which hinders multiple cropping as well as cropping intensity (Fig. 3.2). In addition, local agricultural offices are advising farmers in the *haor* region to harvest their paddy if about 80% of the grains become mature to avoid massive damage from disasters (DAE 2016). However, the land productivity of the region for rice production is higer than the national average level (Fig. 3.3).

Table 3.2 Characteristics of households

| | No. of non-farm households | No. of farm households | Categories of farm households* | | | Agricultural labor household |
			Small	Medium	Large	
Bangladesh	13,512,580 (47.08%)	15,183,183 (52.92%)	12,812,372 (84.38%)	2,136,415 (14.07%)	234,396 (1.54%)	30.82%
Kishoreganj District	289,019 (48.35%)	308,733 (51.65%)	261,308 (84.64%)	41,302 (13.37%)	6,123 (1.98%)	37.50%

*Note Small, medium and large farm families have less than 2.50 acres of land, 2.50–7.50 acres of land, and more than 7.50 acres of land respectively (*Data source* BBS, Yearbook of agricultural statistics 2020)

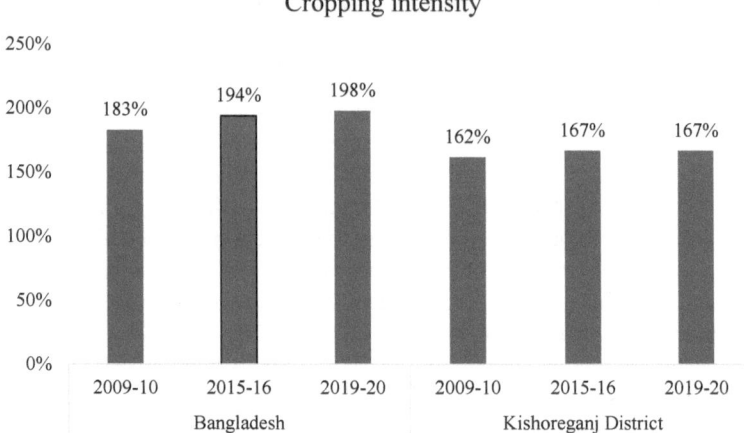

Fig. 3.2 Cropping intensity in recent years (Area in thousand acres) (*Data source* Authors calculation based on the data from Yearbook of agricultural statistics 2010, 2016, 2020, BBS)

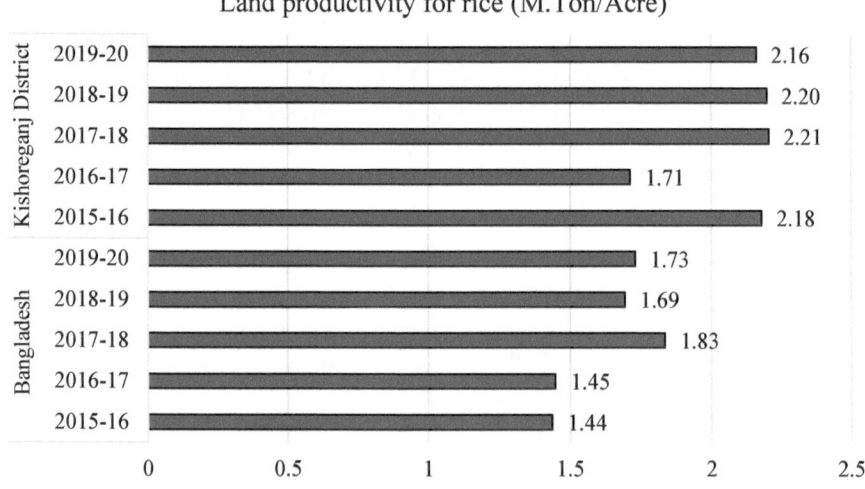

Fig. 3.3 Land productivity in rice production of *haor* region as compared to national average level (*Data source* Authors calculation based on the data from Yearbook of agricultural statistics 2016–2020, BBS)

3.4 Geography and Environment

The *haor* region of Bangladesh is situated in the north-eastern part of the country which is a wetland. The average annual rainfall in the *haor* region is 4130 mm which is almost double the national average (Nowreen et al. 2015). Typically, annual

Table 3.3 Institutional support for disaster risk management in the study area

Indicators	Kishoreganj district	National level
Disaster risk management situation 2009–2014		
Disaster preparedness of households	7.96%	66.54%
Households received financial/rehabilitation support from government/non-government agency during/post disaster period	7.32%	13.87%
Households received loans from post disaster period	12.76%	16.99%

Source BBS (2015)

regular floods from May to October submerge most of the area and turn it into an island. Then human mobility is possible mainly by boat in the region (Gillingham 2016; Rana et al. 2020; Wikipedia). The homestead sites are raised about 1.5 to 2.5 m from the crop field. The level of river water rises up to nearly 0.5 m above the homestead level during rainy season while it goes down up to 8 m below homestead level during dry season. At the same time, people in Kishoreganj district suffer from regular occurrences of river erosion caused by the loss of houses and shelters during the late period of inundation. The situation of Kishoreganj district is characterized by limited livelihood opportunities and frequent crop failure because of early flash flood and poor communication facilities with other parts of the country. Although the target area is one of the disaster-prone areas of the country but disaster recovery support from the government and non-government organizations is also less in the region (Table 3.3).

3.5 Economic Situation

The area faces widespread problems of food insecurity due to disaster risk, limited access to markets and unequal access to resources, etc. (Gillingham 2016). Although the country has significantly reduced the poverty, but it is increased in the study area (Fig. 3.4). As mentioned by Kazal et al. (2017), poverty in the *haor* area is more acute than in other rural areas of Bangladesh. According to BBS (2018), the average annual household income in the district (187,854 BDT) is lower than the national average (202,724 BDT). In the study area, the local agriculture office demonstrates the different technologies for improving farming practices and provides technical support to the farmers.

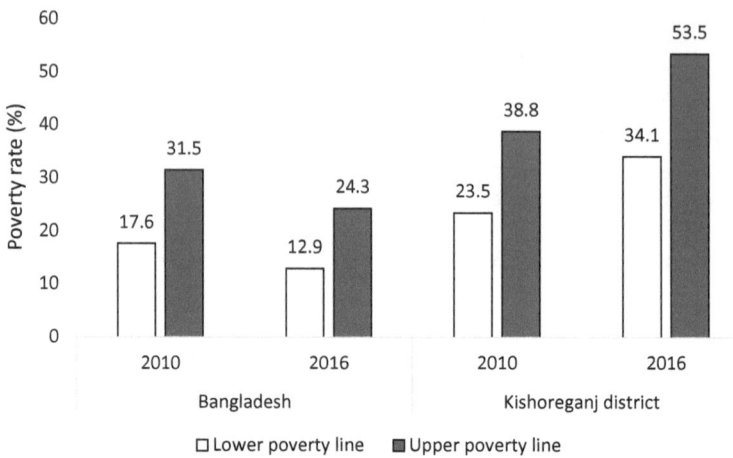

Fig. 3.4 Poverty scenario from 2010 to 2016 (*Data source* BBS, Household income and expenditure survey of Bangladesh, 2010 and 2016)

3.6 Social Structure and Gender Inequality

The literacy rate for both sexes is lower in the region than the national average (Fig. 3.5). Women face greater challenges for access to basic social services. Social tolerance is a barrier for rural women to involve in economic activities (Rana et al. 2021) although initiatives are taken by various stakeholders in the region. The condition of primary social services in Kishoreganj district is poorer than the national level in almost all indicators like education, health, nutrition, sanitation and child protection. The community interaction (bonding social capital) is stronger in the target region than the national average which is considered as an important factor for enhancing regional development.[1] However, bridging and linking type of social capitals are generally weak in the region (Rana et al. 2020). NGOs in the locality provide development programs for woman's awareness toward disaster risk management although their main activity is related to microfinance. One of the NGOs, named BRAC provides education program for the poor children and legal support to the women against violence. The efforts made by different stakeholders are considered have impacts on sustainable development in the region.

---

[1] *Source* Accelerating SDGs in Bangladesh: An Assessment on Coverage of Basic Social Services Database, (http://mscw.bbs.gov.bd/reportmodule/generate_factsheet_district/) Accessed on 21 January 2020.

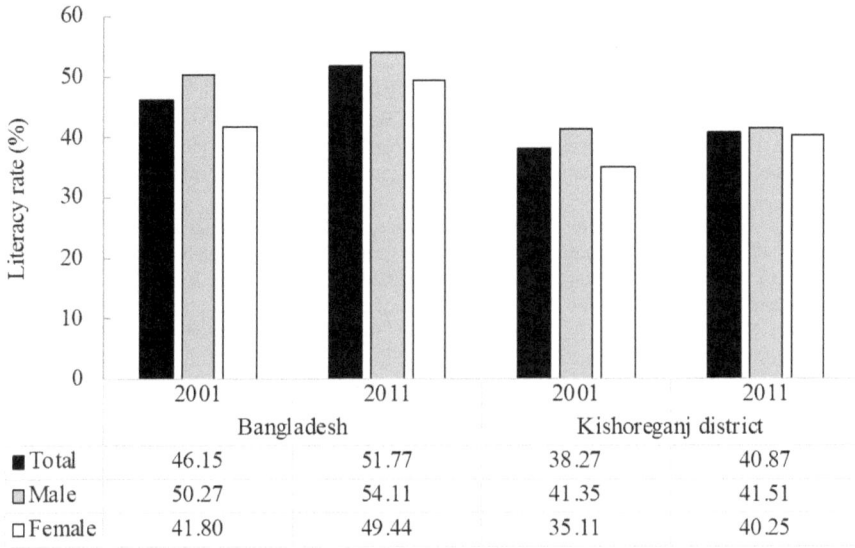

	Bangladesh		Kishoreganj district	
	2001	2011	2001	2011
■ Total	46.15	51.77	38.27	40.87
▢ Male	50.27	54.11	41.35	41.51
▢ Female	41.80	49.44	35.11	40.25

Fig. 3.5 Literacy rate (7 years and above) (*Source* BBS, Population and housing census 2001 & 2011)

3.7 Political Culture and Institutional Collaboration

Local marginalized communities rarely have access to community resources and services due to incompetent management by local elites for their own convenience and institutional weakness. Also, women's participation in local political structures (e.g. union councils) is particularly marginalized due to lack of representation and non-cooperation in institutional structures (Kabir et al. 2018). Moreover, limited transparency and poor institutional collaboration, and low participation of local citizens in the decision-making process hampered the effectiveness of vulnerable group development programs in the region. In addition, CARE Bangladesh pointed out that, there is a need for ongoing programming and a strengthening of the local institutions as well as the collaboration of each other in the region to address the unique environmental vulnerabilities and poverty problems (Gillingham 2016). Although a sound relationship between GO and NGO is very important to face the future challenges of regional socio-economic development of Bangladesh especially in the *haor* region of the country, but there is still a big gap.

References

BBS (2011) Population and housing census, (2011) Bangladesh Bureau of Statistics, Ministry of Planning. Government of the People's Republic of Bangladesh, Dhaka

BBS (2015) Bangladesh disaster-related statistics 2015, climate change and natural disaster perspectives. Bangladesh Bureau of Statistics, Ministry of Planning, Government of the People's Republic of Bangladesh, Dhaka

BBS (2018) Report on agriculture and rural statistics, (2018) Bangladesh Bureau of Statistics, Ministry of Planning. Government of the People's Republic of Bangladesh, Dhaka

DAE (2016) Agricultural extension manual. Department of Agricultural Extension, Ministry of Agriculture, Government of the People's Republic of Bangladesh, Dhaka (in Bengali)

Gillingham S (2016) CARE Bangladesh program strategy: haor region 2015–2020. CARE Bangladesh, Dhaka

Kabir SMS, Aziz MA, Shathi AKMSJ (2018) Women empowerment and governance in Bangladesh. Indian Journal of Women and Social Change 3(1):24–35

Kazal MMH, Rahman S, Hossain MZ (2017) Poverty profiles and coping strategies of the haor (ox-bow lake) households in Bangladesh. Journal of Poverty Alleviation and International Development 8(1):167–191

MoA (2020) Rice harvesting machines has been allotted for the haor region on an emergency basis (in Bengali). Ministry of Agriculture, Government of the People's Republic of Bangladesh. https://moa.gov.bd/site/news/730e50c2-0429-4bf4-9b47-c06b9a500027

Nowreen S et al (2015) Changes of rainfall extremes around the haor basin areas of Bangladesh using multi-member ensemble RCM. Theoritical and Applied Climatology 119:363–377

Rana S, Kiminami L, Furuzawa S (2020) Analysis on the factors affecting farmers' performance in disaster risk management at community level: focusing on a haor locality in Bangladesh. Asia-Pacific Journal of Regional Science 4(3):737–757

Rana S, Kiminami L, Furuzawa S (2021) Social innovation for women's empowerment in disaster risk governance: focusing on common interest groups in the *haor* region of Bangladesh. Studies in Regional Science 51(1):145–155

Chapter 4
Research Framework, Hypotheses and Methods

Abstract For achieving the purpose of this study, the conceptual framework is constructed to describe the relationships among entrepreneurship development, social innovation, socio-cultural changes, and disaster risk management toward sustainable regional development. The following hypotheses are set for verification: "Social and cultural changes through the development of entrepreneurship is necessary for the sustainable regional development (H1)"; "Social capital is important for the development of entrepreneurship, however there is a big gap among different social groups (H1-1)"; "CIG is an effective approach to empower women through socio-political transformation (H2)"; "Farmers' performance in disaster risk management at community level is determined by their socio-economic attributes, social capital, and access to local institutions (H3)". Both qualitative (Trajectory Equifinality Modeling) and quantitative (Structural Equation Modeling) analytical approaches are applied to verify the hypotheses in this study.

Keywords Research framework · Hypotheses · Methods · TEM · SEM

Based on the above-mentioned literature review and existing situation of the aspects of regional characteristics of the *haor* region of Bangladesh, the research framework is constructed as shown in Fig. 4.1 for achieving the purpose of the study. The following hypotheses are set for verification: "Social and cultural changes through the development of entrepreneurship is necessary for the sustainable regional development (H1)"; "Social capital is important for the development of entrepreneurship, however there is a big gap among different social groups (H1-1)"; "CIG is an effective approach to empower women through socio-political transformation (H2)" and "Farmers' performance in disaster risk management at community level is determined by their socio-economic attributes, social capital, and access to local institutions (H3)".

The methodologies used in this research for hypothesis verification are as shown in Table 4.1. First, we will clarify the role of entrepreneurship and social innovation for social and cultural changes in the region using the method of Trajectory Eqiufinality Modeling (TEM) based on the case studies of both male and female entrpreneurs from the target area. As we mentioned earlier that, the development of entrepreneurship is widely recognized as a tool for sustainable development of a

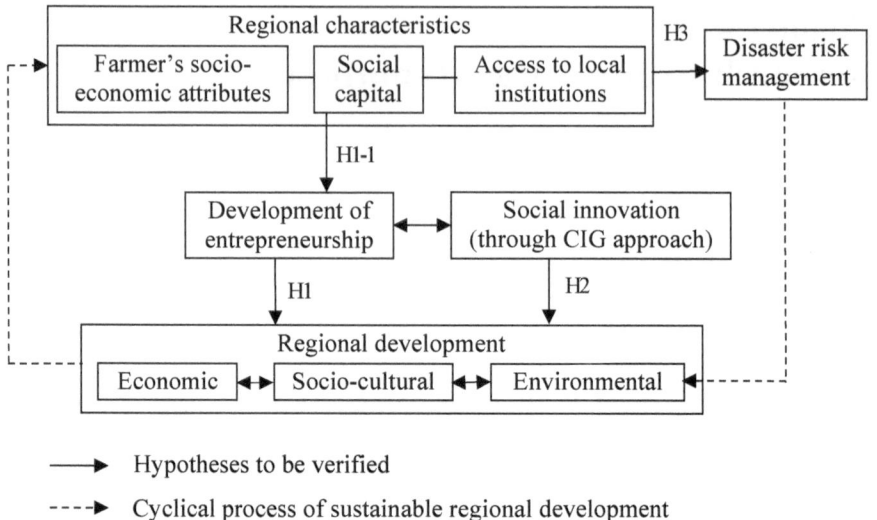

Fig. 4.1 Conceptual framework

region which includes economic, socio-cultural, and environmental development and all these are interconnected with each other. However, the regional charecteristics such as socio-economic situation, social capital, and access to local institutions etc. play a vital role to regulate the entrepreneurial ecosystem in a region (EU, 2014). Moreover, entrepreneurship development and social innovation are interlinked with each other. Secondly, we included the impact analysis of common interest group (CIG) approach on socio-political transformation in the *haor* region of Bangladesh. Thirdly, we will clarify the multi-causal relationships among the factors affecting the performance in DRM using Structural Equation Modeling (SEM) analysis based on the results of questionnaire survey of the farm households of the target area. The *haor* region of Bangladesh is one of the disaster-prone regions of the country. The people of the *haor* region usually live with disasters. In addition, disaster risk management (DRM) is a multidisciplinary approach that has complex relationship with different factors and it is needless to say that without proper DRM, sustainable development in the region is quite impossible.

Trajectory Equifinality Modeling (TEM) is a qualitative research methodology of social and developmental science study which can depict the variation of trajectories of an individual in relation to the society in irreversible time (Sato et al. 2009; Sato and Valsiner 2010). The TEM shows the irreversible time with a horizontal axis but not in a concrete unit, which means reaching similar results while following different paths. It is characterized by the concept of the equal solstice, which is the point of realizing of management decision making of an individual, the branch point where multiple options are prepared for the target and the variety of routes by the equal

Table 4.1 Hypotheses and methods used in the study

Hypotheses	Methods
H1: Social and cultural changes through the development of entrepreneurship is necessary for the sustainable regional development H1-1: Social capital is important for the development of entrepreneurship, however there is a big gap among different social groups	Case study, Trajectory Equifinality Modeling (TEM)
H2: CIG is an effective approach to empower women through socio-political transformation	Key informant interviews (KII), Focus group discussion (FGD)
H3: Farmers' performance in disaster risk management at community level is determined by their socio-economic attributes, social capital, and access to local institutions	Questionnaire survey, Structural Equation Modeling (SEM)

solstice and branch point. In this research, we analyzed the cases of entrepreneurs based on the TEM analytical model shown in Fig. 4.2.

In other words, changes in management decisions that seem to have a significant impact on entrepreneurial development were set as main bifurcation points (BFP). In addition, we consider the socio-economic factors that can influence the management decision as to the social direction (SD) and the social guidance (SG). Since this study focuses on the verification of Hypotheses, the equifinality points (EFP) were not described in detail. As an alternative, we emphasize the next bifurcation point (BFP2) in the business decision making instead of equifinality point (EFP1). In the TEM analysis, we considered the process of entrepreneurial development especially

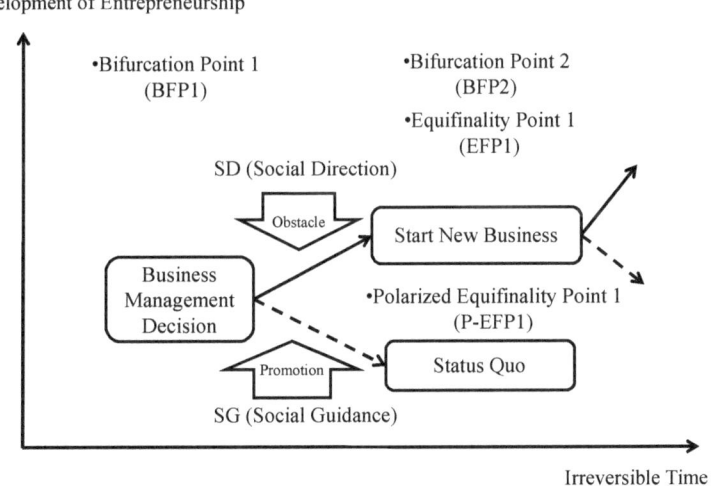

Fig. 4.2 TEM analytical model (*Source* Modified from Kiminami et al. 2020)

for the cases of women entrepreneurs as a means of empowerment. This is because the patterns of social direction (SD) and social guidance (SG) may be different for men and women which are reflected by their images in the socio-cultural structure.

Structural Equation Modeling (SEM) is a multivariate statistical analysis technique that is used to analyze structural relationships among the variables. This analytical technique includes the combination of factor analysis and multiple regression analysis, and it is used to analyze the structural relationship between measured variables and latent constructs (Hox and Bechger 1999). According to Schumacker and Lomax (2010) the reasons for the attractiveness of SEM are as follows. First, for better understanding of complex scientific inquiry using multiple observed variables, SEM is a good approach of analysis to deal with sophisticated theoretical models. Because, basic statistical methods can only utilize a limited numbers of variables. Secondly, SEM provides more accurate analytical measurement score because it explicitly takes measurement error into account during data analysis. Thirdly, SEM has the ability to analyze more advanced theoretical models representing the complex phenomenon of the study. Finally, SEM software programs have become increasingly user-friendly now-a-days.

References

EU (2014) REDI: The regional entrepreneurship and development index-measuring regional entrepreneurship. Final report. Publication office of the European Union, Luxembourg

Hox JJ, Bechger TM (1999) An introduction to structural equation modeling. Family Sci Rev 11:354–373

Kiminami L, Furuzawa S, Kiminami A (2020) Social entrepreneurship and social business associated with multiple functions of urban agriculture in Japan. Asia-Pacific J Regional Sci 4:521–552

Sato T, Hidaka T, Fukuda M (2009) Depicting the dynamics of living the life: the Trajectory Equifinality Model. In: Valsiner J, Molenaar P, Lyra M, Chaudhary N (eds) Dynamic process methodology in the social and developmental sciences. Springer, New York, pp 217–240

Sato T, Valsiner J (2010) Time in life and life in time: between experiencing and accounting. Ritsumeikan J Human Sci 20:79–92

Schumacker RE, Lomax RG (2010) A Beginner's guide to structural equation modelling. Routledge, 3rd edition

Chapter 5
Role of Entrepreneurship in Socio-Cultural Changes

Abstract The purpose of this chapter is to analyze the relationship between entrepreneurship development and its role in socio-cultural changes toward sustainable regional development through the case studies of male and female entrepreneurs in the target region by introducing Trajectory Equifinality Model (TEM) analysis. The results of the TEM analysis clarified that, social and cultural changes in the gender relations are not only necessary for innovation creation, but also for sustainable development. The social capital is important for each case of entrepreneurship development although the quality of social capital varies among the different social groups due to their gender, and socio-political backgrounds. Although entrepreneurial activities have greatly improved the region's economic situation, the performances among social groups are different significantly due to the level and quality of social capital and technology adoption.

Keywords Socio-cultural changes · Entrepreneurs · Social capital · Technology adoption · TEM

5.1 Selection of Entrepreneurship Cases

We selected six cases of entrepreneurs (Table 5.1) including three male and three female cases and conducted the semi-structured interview (see Appendix 5.1) for in-depth investigation. The rationale for the selection of these cases is as follows. First, in the selection of cases of three male entrepreneurship three different categories of farmers (large-scale, small-scale, and landless) are considered, since agriculture is the main livelihood of most of the people in the *haor* region of Bangladesh. Secondly, in the selection of the cases of female entrepreneurs, we obtained initial information from the Upazila Women Affairs Department, Itna, Kishoreganj. According to Global Entrepreneurship Monitor (GEM) Bangladesh report of 2011, there are very few women entrepreneurs in Bangladesh due to a lack of opportunities to work independently in the society (Karim and Hart 2011), and the involvement of women in business activities is also very low in the study area (see Appendix 5.2).

Table 5.1 Description of the case selection for TEM analysis

Entrepreneurship cases	Source of information
Case 1 Mr. AI (Business development of rental service of combine rice harvester)	[Research article] Islam et al. (2019) [Report] MoA (2020) Rice harvesting machines has been allotted for the *haor* region on an emergency basis. Ministry of Agriculture, Government of the People's Republic of Bangladesh. (In Bengali) (https://moa.gov.bd/site/news/730e50c2-0429-4bf4-9b47-c06b9a500027) (Accessed 15 January 2021) [Local Institution] Upazila Agriculture Office, Itna, Kishoreganj, Bangladesh [Semi-structured Interview Survey], February 2021 (Online)
Case 2 Mr. SH (Business development of agricultural inputs)	[Research article] Uddin et al. (2019) [Report] DAE (2020) Report on fertilizer dealers and retailers on 30 June 2020 (in Bengali). Department of Agricultural Extension, Government of the People's Republic of Bangladesh. (http://www.dae.gov.bd/site/page/d3b65ada-99a0-429b-bc86-437dd722ec52) (Accessed 15 January 2021) [Local Institution] Upazila Agriculture Office, Itna, Kishoreganj, Bangladesh [Semi-structured Interview Survey], February 2021 (Online)
Case 3 Mr. ZM (Business development as a retailer)	[Local Institution] Upazila Agriculture Office, Itna, Kishoreganj, Bangladesh [Semi-structured Interview Survey], February 2021 (Online)
Case 4 Ms. SO (Business development of making, designing and selling cloths)	[Local Institution] Upazila Women Affairs Department, Itna, Kishoreganj, Bangladesh [Semi-structured Interview Survey], March 2021 (Online)
Case 5 Ms. RL (Business development of making handicrafts)	
Case 6 Ms. SA (Business development of sewing and making handicrafts)	

Ferdousi and Mahmud (2019) mentioned that 71% of women entrepreneurs in rural Bangladesh conduct their business at home and are involved in making, designing, and selling clothes and handicrafts.

5.2 Case Studies from Male Entrepreneurs

5.2.1 Case 1: Mr. AI (Entrepreneurship and Business Development of Rental Service of Combine Rice Harvester)

The case of Mr. AI was selected for this study because the issue of entrepreneurship development for rental service business of combine rice harvester has appeared as a viable business model in the *haor* region of Bangladesh through enhancing secure rice harvesting and reducing disaster risks (Islam et al. 2019). It has been clarified that farmers would be able to save 40% of their rice harvesting cost through combine harvester rental services (Islam et al. 2019). Moreover, the Government of Bangladesh has recently taken steps to introduce a farm mechanization system for farmers to buy rice combine harvester with the subsidy in the *haor* areas (MoA 2020). Mr. AI is one of the beneficiaries who has received 70% government subsidy on the purchase price of rice combine harvester in the *haor* region.[1] He has created local employment opportunities by promoting farm mechanization and reduced disaster risk in the area through his business model.

Mr. AI is a farmer entrepreneur lives in Itna, Kishoreganj of Bangladesh. He is 52 years old and has a primary level of education (class five). In addition to the rental service of combine rice harvester, Mr. AI launch a farm machinery parts business and inaugurated repair services on other's farm machines e.g. Irrigation pumps to promote farm mechanization in the *haor* area. The results of TEM analysis for Mr. AI is as shown in Fig. 5.1.

Mr. AI had a large-sized family farm (more than 15 acres of land). From his childhood, he was interested in agricultural machinery and realized that it is very difficult to manage large scale farming in the *haor* area without mechanization (SG1). He looked for a solution for solving problems especially while harvesting (unavailability of labor during harvesting and at the same time risk of flooding during harvesting) (SD1). He obtained information about the rice harvesting machine through his elder brother who has good communication with some of the staff members of Bangladesh Agricultural University (BAU). Although Mr. AI did not succeed and faced economic losses due to lack of technical operating skills and the necessary infrastructure such as farm roads, various rivers, delays in getting emergency technical assistance for machines and land use patterns, etc. in the *haor* area (SD2), he did not lose his courage and interest.

In 2013, he bought a mini rice harvester (Table 5.2) and started his rental service business to other farmers in the locality as well. In addition, he learned maintenance and repairing skills of the farm machines by himself (SG3). Moreover, he started a machinery parts business in the area and inaugurated repair services on other's farm machines e.g. irrigation pumps to promote farm mechanization (BFP2). He bought a head-feed combine harvester in 2018 (BFP3) and a full-feed combine harvester in

---

[1] Source: Itna Upazila Agriculture Office, Kishoreganj, Bangladesh.

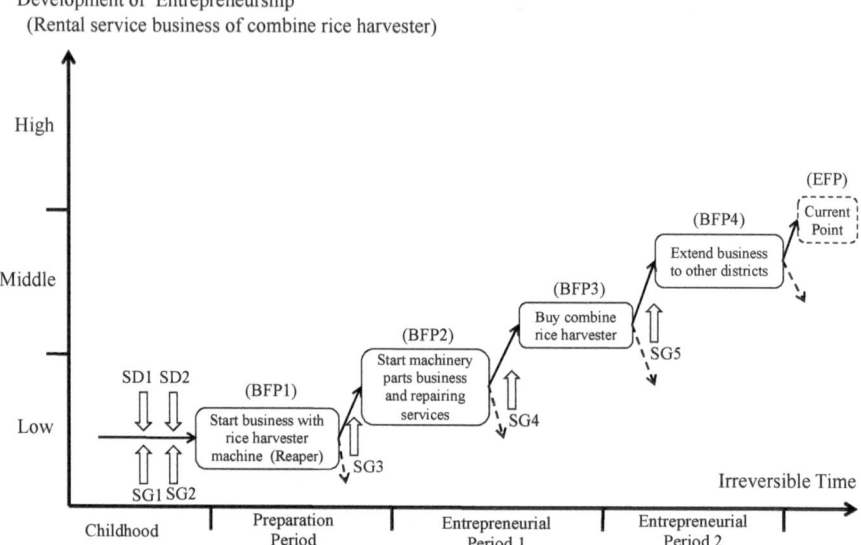

Fig. 5.1 Results of the TEM analysis (Mr. AI)

2019 with the 70% subsidy of the purchase price from the Government of Bangladesh (SG4) for each. Mr. AI mentioned that it is only possible to buy such machines at very high prices with the help of government subsidies, especially in *haor* areas (poor areas). If the plot is small in size, the farmers who own the land pay him 5000–6000 BDT per acre of land for rice harvesting.

Since 2020, he has started to expand his rental service business of rice combine harvester to other districts (BFP4), because the use of harvesting machines is only for a short duration in the *haor* area (around one month; from the middle of April to the middle of May) it could work in other districts in the alternative season. Although there are many challenges such as special boats or trucks required for transporting machines and local assistance in the concerned areas, he mentioned that he has got

Table 5.2 History of purchase of rice harvesting machine by Mr. AI

Sl. no.	Name of harvesting machine	Purchase price (BDT)[a]	Purchase time	Financial support
1	Reaper	250,000/-	2004	Self-fund
2	Mini rice harvester	400,000/-	2012	Self-fund
3	Head-feed combine harvester	18,00,000/-	2018	Received 70% subsidy from the government
4	Full-feed combine harvester	33,50,000/-	2019	

[a]*Note* BDT = Bangladeshi Taka; 1US$ = 84 BDT approximately

the support of the agriculture office, local government office or local elite in the areas (SG5). He currently owns four rice harvesting machines. One is driven by himself, two are driven by his two sons (the school dropped) and the other is driven by a hired labor. Each machine needs another person as a supporting staff during the operation. He is the pioneer of introducing these rice harvesting machines in his area. Now he paid more attention to his rental service business of rice combine harvesters rather than his family farming (EFP). He leased most of his land to other farmers. This year (2021) he planted only 5 acres of land on his own.

The following points are important findings obtained through the case analysis of Mr. AI. First, the entrepreneurship of Mr. AI has been raised along with the development of a new business model of rental service of rice combine harvester in the *haor* area in the processes of trial and error. Secondly, social capital such as good relationships with his family and the human network with the university, and the support from governments (including subsidy) are also important for the business development.

5.2.2 Case 2: Mr. SH (Entrepreneurship and Business Development of Agricultural Inputs Supply)

The case of Mr. SH was selected for this study because of his unique business model for providing agricultural inputs such as pesticides, fertilizers, and fuel for farm-machinery to the farmers in the *haor* area. The results of the TEM analysis for Mr. SH is as shown in Fig. 5.2. Mr. SH is an agricultural input supplier (entrepreneur) lives in Itna, Kishoreganj district of Bangladesh. He is 48 years old and has a lower secondary level of formal education (class six). He had been engaged in family farming with his father since childhood. However, he found that the small-scale family farm was hard to win the competition for survival (SD1). He started the stationery business with a small grocery store in 2003 (BFP1), but he could not make his business profitable because there had been many similar businesses in the area. When he realized that there was a high demand for agricultural inputs but few businesses dealing with (SG1), he took it as a great business opportunity. After applying to the DAE district office for a license to trade in pesticides and received it in 2006, he shifted his former business to the pesticides business (BFP2).

For expanding his business, he also got permission to sell chemical fertilizers as a retailer through institutional support from a local fertilizer dealer and included the selling of fuel (e.g. diesel) for farm-machinery in his business (SG2). In 2011, he successfully got BCIC dealership which is not only an opportunity for conducting business on selling chemical fertilizer, but also a social prestige in his region (BFP3). The important characteristic of Mr. SH's business model is to have good communication with different stakeholders in the area. Due to the expansion of the business, he hired two employees from the region. He received institutional support from GOs

Development of Entrepreneurship
(Agricultural inputs supply)

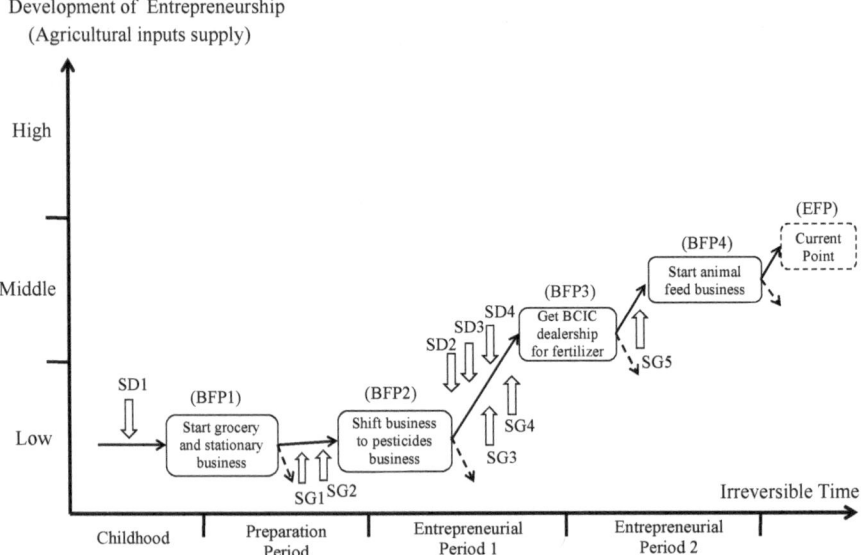

Fig. 5.2 Results of the TEM analysis (Mr. SH)

and NGOs to run the business smoothly. A few years ago, he received credit assistance in his financial crisis from BRAC and POPI, two NGOs in the locality (SG3). On the other hand, he often provides farmers the information about how to select and use pesticides and fertilizers appropriately, and how to apply them at a good timing, etc. He realized that he should also make a contribution to the farming community in the area through his business. For instance, he provided some community services such as providing pesticide spray machines free of cost, and allowing the farmers to pay the purchase price of chemical fertilizers and pesticides after the harvest when they cannot afford them immediately due to poverty or lack of credit access.

However, the main challenges in his business are the difficulty of getting fertilizers timely from buffer godowns or BCIC factories located in adjacent areas due to bureaucratic problem and poor transportation system in the area (SD2). Availability of chemical fertilizers at an appropriate time is very important for farming practices. He also faced instability in his business from the local elites affiliated with the rolling political party and pressured him to change his political affiliation and he did it (SD3). In addition, the demand for agro-chemicals in the region is almost nil as there is almost no agriculture during the monsoon season (May to October) due to water-logging (SD4). Since 2018, he has diversified his business to sell animal feed which has a high demand (BFP4) during the monsoon season in the *haor* region (SG4). He also sells fuel e.g. diesel for boat engines because boats are the only way for people to move and carry goods during the rainy season in the *haor* area.

The following points are important findings obtained through the case analysis of Mr. SH. First, Mr. SH perceived the opportunities for the entrepreneurship and business development of agricultural inputs supply in the *haor* area through self-motivation to reach the equifinality point (EFP). Secondly, social capitals such as a good relationship with different stakeholders and institutional supports are important for business development although socio-political issues hinder the process of business development. Thirdly, in addition to honesty, hardworking and good communication with stakeholders, community contributions through business are also required for being a good entrepreneur in *haor* area.

5.2.3 Case 3: Mr. ZM (Entrepreneurship and Business Development as a Retailer)

The case of Mr. ZM was selected for the study due to he developed a diversified business model as a retailer in the *haor* area. The results of the TEM analysis for Mr. ZM is as shown in Fig. 5.3. Mr. ZM is an entrepreneur (retailer and restaurant owner) who lives in a rural haor area of Itna, Kishoreganj district of Bangladesh. He is 34 years old and has a primary level of formal education (class five). He had been involved in family farming since childhood with his father who was a sharecropper. His father's death in 1999 put his family in financial trouble and he stopped studying (SD1). Then, with the help of friends and relatives (SG1), he started a small business in 2001 with a grocery store in the local market (BFP1) to get out of poverty. He was succeeding in his business step by step due to the good location of his business organization and his sincerity and hardworking qualities. Now he owns two grocery stores in his area. He hired a local employee and his elder son (a seventh-grader) also helped him part-time because of the expansion of his business. However, he had difficulty gaining access to institutional credit assistance (SD2) to expand his business.

Mr. ZM looked for other business opportunities in his area. He included providing phone recharge service on a small scale beside his grocery business which is a most promising business in the *haor* area nowadays. In the future, he wants to be a local agent of phone companies for this business which requires high investment. Since 2018, he has started selling LP gas and gas stoves with a local delivery service (BFP2) due to the increasing demand for changing lifestyles of *haor* people (SG2). He mentioned that it is an emerging and profitable business in the region. Recently, he got an offer from a national company through one of his relatives working for the company, to distribute rice and wheat products, and edible oils in the local market of three adjacent upazilas (sub-districts) as a distributor. But he cannot afford this offer due to a lack of financial capacity. He also mentioned that it is not easy for him to get credit support from a formal financial institution such as a local bank.

He commenced a restaurant business in his area in 2021 (BFP3) when he recognized that many people now travel to the *haor* areas every day from other parts of the country for different purposes due to the improvement of road communication facilities (SG3). So he took it as an opportunity to open a restaurant in his area to serve

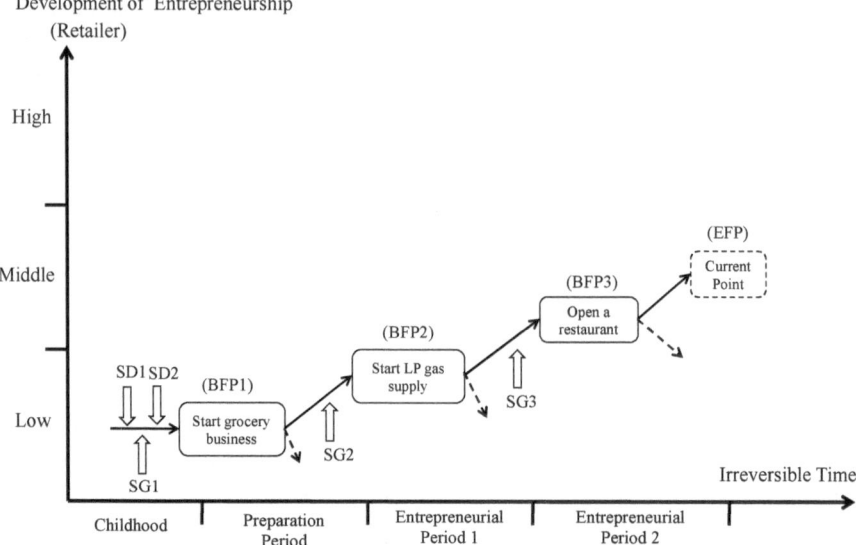

Fig. 5.3 Results of the TEM analysis (Mr. ZM)

food to the visitors and local people as well. Now he is a well-known businessman in the area and is working for expanding the business in the region (EFP). Once there was extreme poverty in his family, now he is happy to have prosperity back in the family through his business model.

The following points are important findings obtained through the case analysis of Mr. ZM. First, Social capital such as a good relationship with friends and relatives is important for his business development. Secondly, the lack of credit access from the local financial institution obstructs entrepreneurial development in the region.

5.2.4   Summary of the TEM Analysis of the Cases of Male Entrepreneurs

Entrepreneurs are embedded in the local community and benefit from different types of path-dependency (agricultural skills, arable land, the tradition of socio-political mobilization, etc.) played a significant role in setting up and developing these initiatives. The findings from the three male case analyses (Table 5.3) provide evidence that the development of entrepreneurship contributes primarily concerning the regional economic growth by promoting the following areas.

Table 5.3 Comparison among three cases of male entrepreneurs

Parameters	Case 1 (Mr. AI)	Case 2 (Mr. SH)	Case 3 (Mr. ZM)
Initial condition	Large-scale family farm (15 acres of land)	Small-scale family farm (5 acres of land)	Landless farmer
Education level	Class five	Class six	Class five
Type of business	Rental service of combine rice harvester	Agricultural inputs supply	Retailer and restaurant owner
Land size	15 acres	5 acres	He has mortgaged 2 acres of land
Motivation to start the business	He realized that it was very difficult to manage large scale farming in the *haor* area without mechanization	Small-scale family farm was hard to win the competition for survival	To get out of extreme poverty
Employment created in the business	8	2	2
Monthly average income (approximately)	187,000 BDT	80,000 BDT	25,000 BDT
Social network	Family, university, local institutions, entrepreneurs (bonding, bridging and linking social capital)		
Empowerment opportunity	Access to local government institutions, finance, extension agencies etc.		
Technology adoption	Hard (Farm machinery)	Hard (Pesticide spray equipment)	Hard (Farm restaurant)
Similarities among the male entrepreneurs	• Social capital is significant for each case of business development • Self-motivation of the entrepreneurs to their business development • Creation of local employment in the region • Contribution to the community through their business activities • Low level of education (class five or six)		
Dissimilarities among the male entrepreneurs	• The initial condition of each entrepreneur was different (Large-scale farmers, small-scale farmers and landless farmers) • The business model of each was different • Institutional support such as access to finance from formal institutions was not equal to the all (this is an indication of corruption)		

All these cases of entrepreneurship have created local employment opportunities and earned more profit to get out of poverty. They contribute to the community such as diversifying farming practices, agricultural inputs supply, reducing the cost of agricultural production, reducing disaster risk, and developing other services to enhance the peoples' quality of life, etc. through their diversified business activities. Moreover, social capital was found significant for each case of business development.

Social capital is defined as "the features of social organization, such as trust, norms, and networks that can improve the efficiency of the society by facilitating coordinated actions" (Putnam 1993, p. 167). In the study area, bonding social capital is strong for all the cases of entrepreneurs but other forms of social capital such as bridging and linking social capital are different for them due to their socio-economic and political background. It is clarified that their businesses contribute to regional economic growth but there are limitations to addressing the socio-cultural transformation. For example, as they employed only male workers in their business and accesses to the finance and community resources are not equal even among the male entrepreneurs.

5.3 Case Studies from Female Entrepreneurs

5.3.1 Case 4: Ms. SO (Entrepreneurship and Business Development of Embroidery and Tailoring)

The case of Ms. SO was selected for this study due to she developed a business model in the rural *haor* area of Bangladesh and inspired other women to become independent through skill development. The results of the TEM analysis for Ms. SO is as shown in Fig. 5.4. Although the government has taken steps to expand education for all, especially girls, it is not easy to pursue higher education for girls in rural *haor* areas due to various socio-cultural barriers such as they are always inspired to get married early, leave school and stay at home. Ms. SO is one of them who got married in the early stage of her life even before the Secondary School Certificate (SSC) examination in 2007 (SD1). However, she has completed a master's degree in history (BFP1) in 2014 from National University with the support of her husband and other members of her father-in-law's house (SG1).

After completing graduation from National University, she started a job in BRAC (an NGO) education program as a project officer in Itna, Kishoreganj (BFP2) in 2013. After a certain period, she was transferred to a nearby sub-district namely Karimganj (SD2). Thereafter, she faced difficulties continuing her job and family maintenance due to the poor transportation system in the area. Moreover, her daughter became ill and needed extra care. Eventually, she resigned from the job. And then she had been thinking to start a business that could be managed at home. In 2015, she participated in a 7-day training program on embroidery and sewing work organized by Itna LGED (SG2). She was inspired by the training and also got some insight to become an independent entrepreneur during her job period in BRAC. She started her embroidery and sewing business (to make designs in men's and women's wear, bed sheets, etc. with colored yarns) in 2016 with her savings (BFP3). She received support from her family members to start the new business (SG3) except for financial support (SD3). Whatever she earns from her business she invests in expanding her

Development of Entrepreneurship
(Embroidery and Tailoring)

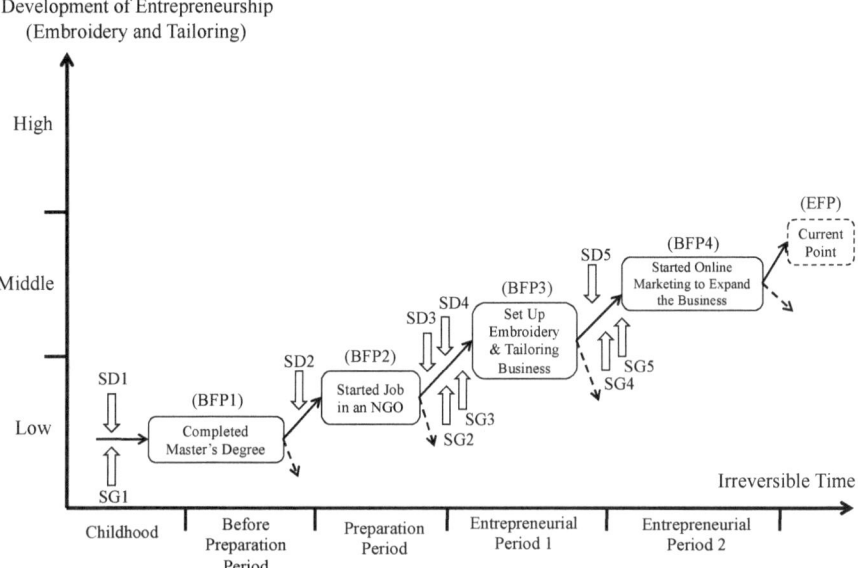

Fig. 5.4 Results of the TEM analysis (Ms. SO)

business. She mentioned that she would need credit support in the future to expand her business. She had to face the issue of social tolerance by being criticized badly for her work as many other women in the community during her initial stage of the business (SD4). However, she did not lose her courage and interest. But now, the people's cognition has been started to change to her entrepreneurial activity.

In 2019, she participated in a 3-month entrepreneurial skills development training program for women organized by the Itna Upazila Women's Affairs Department. She learned different aspects of entrepreneurial development from the training including the development of marketing channels (SG4). Then, she started online marketing about the products of her business through her Facebook account (BFP4). This helps her to become more familiar with her business and start getting orders from customers of other regions. She sells the products at a slightly lower price than other similar businesses because she can reduce production costs by using available resources such as cheap labor (SG5). She also gets orders from other traders of different regions before various festivals like Eid, Puja, Valentine's Day, etc. as the demand for these products at festivals is much higher than usual time. However, she faced difficulties in delivering the goods to the customers as there is no courier service in her area (Itna). She has to travel at least 15 km to a nearby sub-district (Mithamin) to deliver the goods through the official courier service which causes extra costs for her due to poor transportation systems (SD5).

She trained 20 women on embroidery and tailoring from the locality who are working part-time in her business. She provides the raw materials to them and they do the work in their home at their convenience and return the final products to her. Part-time workers include different categories of women such as housewives, students, etc. She mentioned that it is an additional source of income for women in the area which helps them to be economically empowered. However, she would also like to employ men in her business in the future but she feels more comfortable working with women. Her husband went to the Maldives in 2015 as an immigrant to earn more and will return to his home in 2022. She wants to establish a showroom in her locality and her husband will operate it if he wants, although he has a family farm. Now she feels that, she has overcome the initial contextual barriers, so it is time to concentrate on expanding the business (EFP). In addition, she made all the decisions about business management by herself. She inspires other women of the locality to be independent and resourceful through skill development. At this stage, she is well known in her area as a female entrepreneur. She strongly argued that; if women get an equal opportunity they will do much better.

The following conclusions are drawn from the case analysis of Ms. SO. First, social capital is important for her to pursue higher education and develop a business model although she has limited access to the family resources. Secondly, she has trained the unemployed women of the locality by providing training and employing them as part-time workers in her business. Thirdly, the perception of people in the community about women's involvement in economic activities has changed due to her entrepreneurial development. This is a positive social change that has started in the area.

5.3.2 Case 5: Ms. RL (Entrepreneurship and Business Development of Making Handicrafts)

The case of Ms. RL was selected for this study due to she developed a business model of making handicrafts in the rural *haor* area of Bangladesh. The result of the TEM analysis for Ms. R L is as shown in Fig. 5.5. Ms. RL is a female entrepreneur who lives in the rural *haor* area of the Kishoreganj district in Bangladesh. She is 22 years old. In 2021, she passed the Higher Secondary Certificate (HSC) examination. Now she is preparing for the admission of Bachelor of Arts (BA) at a college in her area. Her father had a small grocery business in the local market and her mother has a sewing business in their home. She has two sisters. Her older sister was married in the same area and her younger sister is a tenth grader. A few years ago, her father had a road accident and broke his left arm which causes a huge medical expense for her family, and unfortunately, he was unable to work.

She has had the experience of working with her mother (BFP1) and has always been searching for how to become self-reliant (SG1) since childhood. When she was

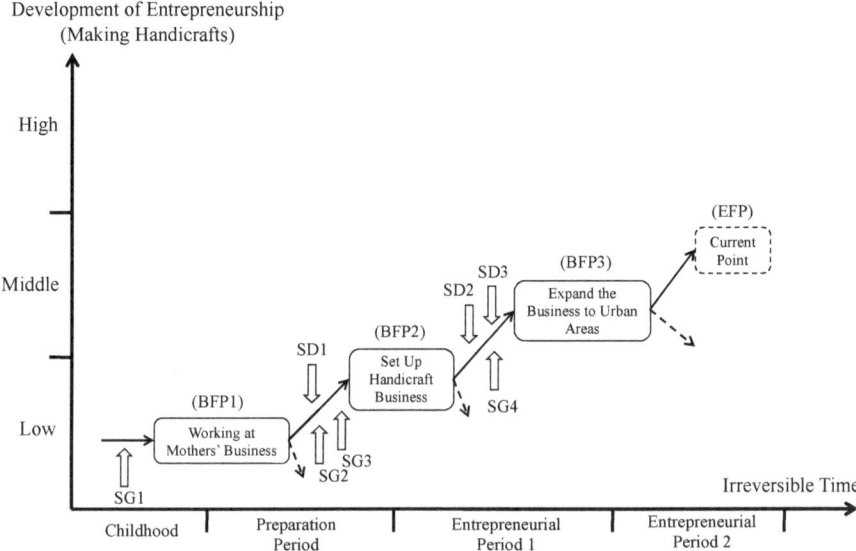

Fig. 5.5 Results of the TEM analysis (Ms. RL)

a student of class 9 (2017), she got the business idea of making handicrafts (crystal showpieces) from one of her women neighbors who runs this business. After receiving a hands-on-training from her neighbor (SG2), she also started a new business of making handicrafts (crystal showpieces) in 2018 (BFP2). Since her mother owns a sewing business, local women usually come to her mother to make clothes. She took it as an opportunity to introduce her new business concept to the local community and considered them as her primary customers. She made various things like key rings, pen cases, flowerpots, bags, doorbells, wall mats, etc. She invested her small saving in her business from the scholarship received from the government as a student. She intends to continue this business after completing her education.

In 2019, she noticed a circular about a 3-month training program for local women organized by the Itna Upazila Women's Affairs Department for helping them become entrepreneurs. She decided to join the training program. She was inspired by the training program and learned different skills for her business development (SG3). At the end of the training she also received BDT 6000 as remuneration. So, she invested this money as her business capital. She started expanding her business in 2019 (BFP3). However, she realized that the area she is living is a poor and unsuitable one for running this business on a large-scale (SD1), so she considered expanding her business to other places, especially in urban areas. One of her uncles lives in Dhaka, the capital city of Bangladesh (SG4). She sent products to him, he delivered them to different stores and she received the money through bkash (a mobile money transfer system in Bangladesh).

However, sending goods through the official courier service is not so easy for her as there is no courier service agent in her area (Itna). She had to go to a nearby sub-district (Mithamin) at least 15 km away to deliver goods through the official courier service, which cost her extra due to poor infrastructure (SD2). She mentioned that access to finance is a big problem for doing business, although there are few NGOs providing loans for small businesses with a high interest rate (SD3). She has a plan to set-up a business showroom in the nearest city area by herself in the future (EFP). She had to manage everything from buying raw materials and selling the products by herself. She had to travel many places alone due to the sickness of her father, the only male member in her family. She feels the business environment is not women-friendly. Sometimes she was not being treated well by other businessmen i.e. and even felt insecure when traveling alone. She said that "I do my business at home. I earn an average of BDT 10,000 per month. My educational expenses have been met up and I am contributing to the family. I feel self-dependent."

The following conclusions are drawn from the case analysis of Ms. RL. First, self-motivation and social capital are important for the development of her business. Secondly, the harsh realities might teach her to become self-independent. It is usually rare in rural areas of Bangladesh, especially for girls, to conduct a business activity along with their studies.

5.3.3   Case 6: Ms. SA (Entrepreneurship and Business Development of Sewing and Making Handicrafts)

The case of Ms. SA was selected for this research due to she developed a business model of sewing and making handicrafts in the rural *haor* area of Bangladesh. The result of the TEM analysis for Ms. SA is as shown in Fig. 5.6. Ms. SA is a female entrepreneur who lives in a rural *haor* area of Itna in Kishoreganj district of Bangladesh. She is 20 years old. In 2020, she passed the Secondary School Certificate (SSC) examination. She is now a Class XI (HSC) student at a local women's college. She has four members in her family, her mother, father, elder brother, and herself. Her mother is a housewife. Her father is a local trader of fish and rice. Her elder brother is serving in the Bangladesh Army.

She started her sewing business in 2017 (BFP1). Ever since childhood, she has been looking for ways to become self-sufficient (SG1). When she was a junior high school student, she started her sewing business after buying a sewing machine through her scholarship money and basic training from a local businessman (SG2). Although her father and elder brother discouraged her from running this type of business instead of concentrating on studies at the beginning (SD1), they finally helped her because of her strong motivation.

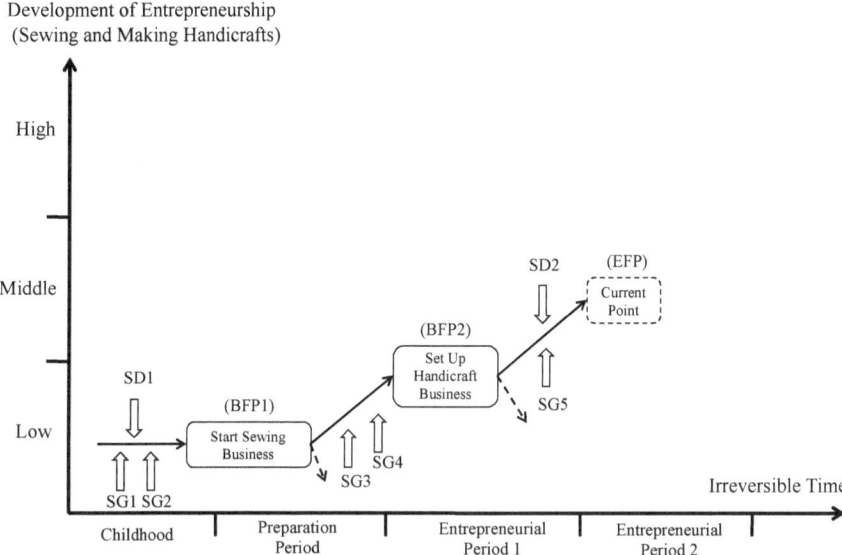

Fig. 5.6 Results of the TEM analysis (Ms. SA)

In 2019, Ms. SA found an opportunity to receive training for three months on entrepreneurial skill development under the Itna Upazila Women's Affairs Department. She learned the necessary skills of making handicrafts with crystal stone, designed cloth, candle, and conducting business activities from the training (SG3). At the end of the training, she also received BDT 6,000 as remuneration. So, she invested this money as the capital of her business. She started a new business of making handicrafts in 2019 (BFP2). She started to make different handicrafts with crystal, designed cloths and candle such as key rings, tissue box, pen cases, flowerpots, apple tree, orange tree, doorbells, wall mats, boat etc. However, she realized that expanding this new business in the region (rural haor area) is a challenge, as these products are consumed only by the relatively wealthy and luxurious people. However, her business in the area is expanding day by day. Her friends are now informed about her business; they also praised her for doing such business. She mentioned that to run a small business at this stage, large sums of money are not required. She said that "I have the skill to make handicrafts, it is my asset". Whatever she earned from her sewing business she invested in her new business (SG4). But in the future, she will need financial support from formal financial institutions to conduct the business on a large-scale, which may not be so easy to access for her (SD2).

She wants to complete her higher studies from university and will continue her business at the same time. She believed that her business would not hinder her studies because she always tried to manage her time properly (SG5). In the future, she wants to do business on a large-scale and convert it into an organizational format (EFP). She has a strong motivation to do something that helps her become self-reliant. "I do my business and earn an average of BDT 8,000 a month. I don't have to tell my family members to pay me for my personal purposes" she said.

The following conclusions have been drawn from the analysis of the case of Ms. SA. First, a strong motivation to become self-reliant is important for the development of her business. Secondly, access to higher education and training is also important for the development of women's entrepreneurship in the region.

5.3.4   Summary of the TEM Analysis of the Cases of Female Entrepreneurs

Although, women's access to higher education is thought to contribute to their empowerment such as (i) increasing their social position (ii) changing their cognition on themselves and (iii) creating higher business sense at different levels through a deeper understanding of the society and broadens their perspectives, etc. women in the selected cases pursuing higher education (Table 5.4) might be special examples, because women are generally considered to be physically, mentally, and intellectually weak in the society of rural Bangladesh (Kabir et al. 2018).

Although there are no legal barriers for women to become entrepreneurs, there are socio-cultural barriers from Bangladesh's perspective. The gender statistics report of Bangladesh 2018 stated that there is no significant relationship between education and women's freedom of movement (BBS 2018). Nouri (2021) clarified that the escalation of commitment of women entrepreneurs is affected by their bitter memories of previous failure, overconfidence, and familial pressure. According to the GEM Bangladesh report of 2011, there are very few women entrepreneurs in Bangladesh due to a lack of opportunities to work independently in society (Karim and Hart 2011). In this situation, joint ventures of entrepreneurial development regardless of gender can minimize this problem.

5.4   Comparison of Male and Female Cases of Entrepreneurs

The results of the TEM analysis of the above cases of entrepreneurs in the study area clarified that although both male and female entrepreneurs have high motivation, their performances are seemed to be different. The entrepreneurial activities of women are confined in the household chores and constrained by the existing social

Table 5.4 Comparison among the three cases of female entrepreneurs

Parameters	Case 4 (Ms. SO)	Case 5 (Ms. RL)	Case 6 (Ms. SA)
Initial condition	She lost her job due to family problems and try to find the alternative ways for income earning	No male earning members in the family	From school life she wanted to earn to support herself
Education level	Masters degree	HSC	SSC
Type of business	Making, designing and selling cloths	Making handicrafts	Sewing and making handicrafts
Land size	No own land	No own land	No own land
Motivation to start the business	To become self-reliant	Self-motivation as well as survival fittest	To become self-reliant
Employment created in the business	20 part-time (for women only)	Self employment	Self employment
Monthly average income (approximately)	15,000 BDT	10,000 BDT	8,000 BDT
Social network	Confined in the family and similar female groups in the community (bonding social capital)		
Empowerment opportunity	Access to support services (e.g. education and training) but less access to finance and asset ownership (e.g. land)		
Technology adoption	Soft (digital marketing through social media)		
Similarities among the female entrepreneurs	• Social capital is significant for each case of the business development • Their entrepreneurial activities are constrained by existing social institutions • Self-motivation of the entrepreneurs to their business development • Nature of business is confined to the household chores • Higher level of education		
Dissimilarities among the female entrepreneurs	• Family structure is different such as married or unmarried		

institutions, and the resistance from traditional male dominance, etc. (Table 5.5). Our findings supported the indication by Neumeyer et al. (2019a, b). The women entrepreneurs in the region have taken advantage of digital technology adoption in their entrepreneurial operations due to the young age, higher level of literacy, and lower level of social network as compared to the male entrepreneurs. They try to find their own alternative ways to build network using social media. Along with entrepreneurial spirit, creating equal opportunities for women in rural haor areas of Bangladesh is also important for their entrepreneurial development as well as empowerment. Therefore, social entrepreneurships through promoting social and cultural changes could solve or mitigate potential problems embedded in the existing social structure in more socially inclusive and context-responsive ways.

Table 5.5 Differences between male and female cases of entrepreneurs

Indicators	Male entrepreneurs	Female entrepreneurs
Level of education	Low	High
Access to household income and assets (e.g. land)	Controlled by them	No or limited access
Social capital	High (bonding, bridging and linking social capital)	Usually confined in the family and similar female groups in the community (bonding social capital)
Nature of business activities	Business development in the community	Limited to household chores
Income	High	Low
Access to local institutions	Male dominance	Limited access
Business environment in the locality	Male dominance	Not women friendly
Social entrepreneurship	Low	Middle

In addition, the women's access to productive resources such as land in the region is lower than the national average (Table 5.6). It is also related to credit access from formal financial institutions. Therefore, sufficient conditions for empowerment are not only the women's access to higher education but also require socio-political transformation to change the existing social structure in the region including the gender gap (Rana et al. 2021). The results of the TEM analysis of the above three cases of female entrepreneurs showed evidence of cultural changes (changes in the cognition toward women's role in the family and community) compared to the cases of male entrepreneurs.

Table 5.6 Ownership of land by female members of the households

Area	Total no of households	No of HHs having own land of women members	% of HHs having own land of women members
Kishoreganj district	568,038	133,321	23.47%
National level (Bangladesh)	27,436,920	8,790,973	32.04%

Data source BBS (2018), Report on agriculture and rural statistics 2018

Appendix 5.1: Semi-structured Interview Guide for the Research on Entrepreneurship and Regional Development: Case Study from a *Haor* Region of Bangladesh

The following interview questions will be used to guide the interview. However, the questions may change as the study progresses depending on the situation.

Basic information about the participant:

Name:

Age:

Sex:

- Male
- Female

Level of formal education:

- Primary (Class 1–5)
- Secondary (Class 6–10)
- Higher secondary (Class 11–12)
- Bachelor or above

Family structure:

- Regular family (i.e. husband, wife, children)
- Female headed household
- Widow
- Single (Not married)

Family members: ....

Childhood memories:

- What are the interesting things?
- What kind of lessons did you learn from your parents?

Formal education in school:

- How much formal education did you receive?
- From which school?
- Why this school?
- What kind of education?
- Why did you stop studying at this stage?

Access to social services and household and community resources:

- How much can you access to the household (e.g. land, income) and community resources (e.g. market) and decision making?

- Do you have access to local institutions such as local government, financial institutions (banks), health services etc.?

Business experience:

- Why did you decide to start the business?
- Do you memorize the moment you wanted to start the business or to become an entrepreneur?
- What type of support did you receive to start your business enterprise? From whom?
- What kinds of obstacle did you face to start your business or to continue your business? How did you overcome these barriers?
- What kinds of skill do you think are required to be a successful entrepreneur? How do you learn these skills?
- What are the motivations for you to become an entrepreneur?
- What kinds of support do you think you need to run your business? (from your experience)
- What are the advantages do you enjoy for being an entrepreneur?
- What are your future prospects as an entrepreneur?

Contribution to the community:

- Have you made contribution to your community since have being an entrepreneur? (in disaster risk governance, poverty reduction, gender equality)
- What are the suggestions would you provide to women who want to start business?

Self-identification:

- How do you identify yourself?

1. .
2. .

Appendix 5.2: Women's Engagement in Business Activities in Bangladesh

See Tables 5.7 and 5.8.

Table 5.7 Total Persons Engagement (TPE) in business enterprises by sex

Year	Kishoreganj district					National level (Bangladesh)				
	No. of enterprises	TPE	Female	Male	% of Female engagement (%)	No. of Enterprises	TPE	Female	Male	% of Female Engagement
2001 and 2003	59,859	147,905	8,880	139,025	**6.00**	3,708,152	11,270,422	1,229,413	10,041,009	**10.90%**
2013	150,946	302,467	44,904	257,563	**14.84**	7,818,565	24,500,850	4,051,718	20,449,132	**16.53%**

Data source BBS (2015), Economic census 2013

Table 5.8 Head of business enterprises by sex

Year	Kishoreganj district				National level (Bangladesh)			
	No. of enterprises	Male headed	Female headed	% of Female headed enterprises (5)	No. of enterprises	Male headed	Female headed	% of Female headed enterprises
2013	150,946	129,858	21,088	13.97	7,818,565	7,255,197	563,368	7.20%

Data source BBS (2015), Economic Census 2013

References

BBS (2015) Economic census 2013 Bangladesh Bureau of Statistics, Ministry of Planning, Government of the People's Republic of Bangladesh, Dhaka

BBS (2018) Report on agriculture and rural statistics 2018 Bangladesh Bureau of Statistics, Ministry of Planning, Government of the People's Republic of Bangladesh, Dhaka

DAE (2020) Report on fertilizer dealers and retailers on 30 June 2020 (in Bengali). Department of Agricultural Extension, Government of the People's Republic of Bangladesh. http://www.dae.gov.bd/site/page/d3b65ada-99a0-429b-bc86-437dd722ec52

Ferdousi F, Mahmud P (2019) Role of social business in women entrepreneurship development in Bangladesh: perspectives from nobinudyokta projects of Grameen Telecom Trust. J Glob Entrep Res 9:58

Islam AKMS, Bhuiyan MGK, Kamruzzaman M, Alam MA, Rahman MA (2019) Custom hire service business of rice combine harvester in haorbasin of Bangladesh. Bangladesh Rice J 23(2):65–75

Kabir SMS, Aziz MA, Shathi AKMSJ (2018) Women empowerment and governance in Bangladesh. Indian J Women Soc Change 3(1):24–35

Karim MS, Hart M (2011) Bangladesh 2011 monitoring report. Global Entrepreneurship Research Association, UK

MoA (2020) Rice harvesting machines has been allotted for the haor region on an emergency basis (in Bengali). Ministry of Agriculture, Government of the People's Republic of Bangladesh. https://moa.gov.bd/site/news/730e50c2-0429-4bf4-9b47-c06b9a500027

Neumeyer X, Santos SC, Caetano A, Kalbfleisch P (2019a) Entrepreneurship ecosystems and women entrepreneurs: a social capital and network approach. Small Bus Econ 53(2):475–489

Neumeyer X, Santos SC, Morris MH (2019b) Who is left out: exploring social boundaries in entrepreneurial ecosystems. J Technol Transf 44(2):462–484

Nouri P (2021) That's why they did not let it go: exploring the roots of women entrepreneurs' escalation of commitment. J Entrepr Emerg Econ 13(2):213–230

Putnam RD (1993) Making democracy work: civic traditions in modern Italy. Princeton University Press, Princeton

Rana S, Kiminami L, Furuzawa S (2021) Social innovation for women's empowerment in disaster risk governance: focusing on common interest groups in the *haor* region of Bangladesh. Stud Reg Sci 51(1):145–155

Uddin MT, Hossain N, Dhar AR (2019) Business prospects and challenges in haor areas of Bangladesh. J Bangladesh Agricult Univ 17(1):65–72

Chapter 6
Socio-Political Transformation Through the CIG Approach

Abstract In this chapter, we conduct key informant interviews (KII), a case study, and focus group discussions (FGD) with various stakeholders of the target region to verify the following hypothesis "CIG is an effective approach to empower women through socio-political transformation" (H2). The results clarified that socio-political transformation for the empowerment of women through the CIG approach is weak in the region even though change has begun. Thus, policy implications based on the analytical results suggest that the government should pay more attention to social innovation to reduce poverty and empower women in disaster risk governance while introducing a CIG approach.

Keywords Socio-political transformation · CIG · *Haor* region · Bangladesh

6.1 Interviews with Key Informants

The purpose of this chapter is to clarify the role of social innovation for women's empowerment in disaster risk governance by focusing on common interest groups (CIG) in the target region. Primary qualitative data was collected through key informant interviews (KIIs), case study and focus group discussions (FGDs) during August 2019. Key informants are the representative of different parts of the community (see Appendix 6.1). A key informant interview (KII) was done with local agriculture officer and officer in charge of a local branch of a national NGO named BRAC; two focus group discussions (FGDs) were conducted with one female CIG and with union council[1] members. Interviews with female and male CIG leaders, local agricultural officer, and female and male union council members (see Appendix 6.2) were

---

[1] The union council is the lowest unit of local government in rural areas of Bangladesh. Union council consists of total 13 members (a chairman and twelve members including three reserved positions for women).

updated in September and October 2020. The social impacts[2] of CIG approach in the study area are explained and the summary of qualitative interviews with different stakeholders are described in Table 6.1.

6.2 Satisfaction of Needs

The majority of the inhabitants of the *haor* region depend mainly on agriculture for their livelihood. A common feature of *haor* agriculture is that the land is submerged for about half a year. Therefore, farmers in the *haor* region get almost half of the opportunity to cultivate their land as compared to other regions of the country. Moreover, their crop harvest is at risk of flood. Farmers have received training on advanced farming practices such as modern paddy cultivation, homestead vegetable gardening,crop diversification and floating vegetable cultivation etc. through the establishment of CIGs, which opens a window for the farming community to acquire improved farming knowledge and technology and access to local institutions.

6.3 Changing Social Relations

The sustainability of agriculture in the *haor* region of Bangladesh largely relies on the efficiency and effectiveness of agricultural extension services, farmers' market connectivity, farmers' access to new farming knowledge and technology, and farmers' entrepreneurship development. NATP emphasizes participatory and knowledge-based agricultural extension services through social mobilization of farmers through the organization of CIGs at the grass-root level for effective transfer of new farming technologies. It has established a platform for increasing farmers' leadership ability, mutual trust (social capital) and self-contributory funds, etc. through the formation of CIGs. It is enhancing the linking social capital through the active interaction between farmers groups and local agriculture office, financial institutions (bank) and, other local institutes. There is a scope of micro-power dynamics and democratic accountability of the executive members of the CIGs. These activities include the development of gender-sensitive local organizations (female CIGs) that can increase the empowerment of women in rural communities. As a result, female CIG farmers become aware of their potential and contribute to the family and community due to working in a group and increased institutional access. In addition, CIG members become more aware of the disaster risks in the locality by exchanging information on disasters with each other in regular meetings, including the cultivation of short-duration rice varieties. Therefore, the CIG approach is facilitating the conditions

---

[2] The social impact is the significant change in people and communities as a result of an activity, a project, a program or a policy (Source: https://www.goodfinance.org.uk/latest/post/blog/social-impact-what-it-how-do-i-measure-it) (Accessed: 23 November 2020).

Table 6.1 Summary of the results of qualitative interviews

Important codes	Respondents				
Effects of CIG approach on women's empowerment in poverty reduction and in disaster risk governance	Female CIG leader	Male CIG leader	Local agriculture officer	Female union council member	Male union council member
1. Poverty reduction					
(a) Increased agricultural production and income	Yes	Yes	Yes	Yes	Yes
(b) Farmers' entrepreneurship development	Yes	Yes	Yes	Yes	Yes
2. Women empowerment					
(a) Leadership development	Yes	Yes	Yes	Yes	Yes
(b) More contribution to family and community	Yes	Yes	Yes	Yes	Yes
(c) Strengthening role in decision making at household/community level	Yes	–	Yes	Yes	Yes
3. Disaster risk governance					
(a) Increased organizational collaboration (GOs, NGOs, CBOs, Business organizations etc.) in the locality	Yes (a little)	Yes (a little)	A little	A little	A little
(b) Increased women representation at community level	Yes	Yes	Yes	Yes[a]	Yes
(c) Information sharing (risk communicator)	Yes	Yes	Yes	Yes	Yes
(d) Mechanized farming reduces disaster risks	–	Yes	Yes	–	Yes
4. Opportunities					
(a) Increased access to newknowledge and technology	Yes	Yes	Yes	Yes	Yes
(b) Tourism in the *haor* area	–	–	–	–	Yes

(continued)

Table 6.1 (continued)

Important codes	Respondents				
Effects of CIG approach on women's empowerment in poverty reduction and in disaster risk governance	Female CIG leader	Male CIG leader	Local agriculture officer	Female union council member	Male union council member
5. Barriers/challenges					
(a) Cultural norms	Yes	Yes	Yes	–	Yes
(b) Access to local market	Yes	Yes	Yes	Yes	–
(c) Female CIGs own farm machines but not operate themselves	–	–	Yes	–	–
(d) Land use pattern	–	Yes	Yes	–	Yes
(e) Local political influence	–	Yes	Yes	Yes	–
(f) Group management	Sometimes	Sometimes	Yes	–	–
(g) Corruption/injustice	–	–	–	Yes	–

[a]Women representative should be 30% at each committee, but in reality, it is very less

for innovation creation through providing a learning environment for the farmers to have a passion to improve farming knowledge and to have confidence in agricultural production.

6.4   Farmers' Entrepreneurship Development and Collective Empowerment

The CIGs formed a self-contributory fund and received grants from the agricultural innovation fund (AIF) of the Ministry of Agriculture to buy farm machineries. There is an opportunity to create a new market for spreading agricultural machineries in the *haor* region. The CIG approach has taken this opportunity to support the farmers' group to adopt farm machineries. These farm mechanization practices not only increase the farm productivity but also reduce the disaster risk in the *haor* area. New crops especially vegetables are now cultivated and floating gardening has been introduced in the region as an innovative farming technique through the CIG approach. The CIGs create a platform for rural farmers to discuss local rights in the socio-political context and some more opportunities as co-production locally.

Collective empowerment is an important dimension of social innovation which requires collective cognitive change through socio-political transformation (Moulaert and MacCallum 2019). Women in the study area are officially recognized as female farmers through the CIG approach. They become aware of their potentials and contributions to family and community as well due to working in female CIGs. The CIGs have enhanced the collective capacity of the rural farming communities by accessing advanced agricultural knowledge and technology through training and demonstration. However, the entrepreneurial development of women farmers in the region is still very limited.

Therefore, the hypothesis 2 "CIG is an effective approach to empower women through socio-political transformation (H2)" has been insufficiently verified due to its weak impacts on socio-political transformation in the study region.

6.5 Obstacles for Socio-Political Transformation

However, there are different barriers for socio-political transformation especially for women empowerment in the *haor* region although the CIG approach has the possibility of cognitive change in some aspects (Fig. 6.1) such as the cultural norms related to gender issues, local political influences, and no or limited access to the local market for women. Women's participation at the community level is increasing over time, but their representation is not being properly valued. For example, under a quarter system for women (at least 30%), there are still very few female members in every working committee in the union council.

Furthermore, as for owning farm machines, female CIGs do not run these machines themselves but rent them as a source of income due to two barriers to operating the farm machinery, one is technical skill and another is social tolerance. In a certain

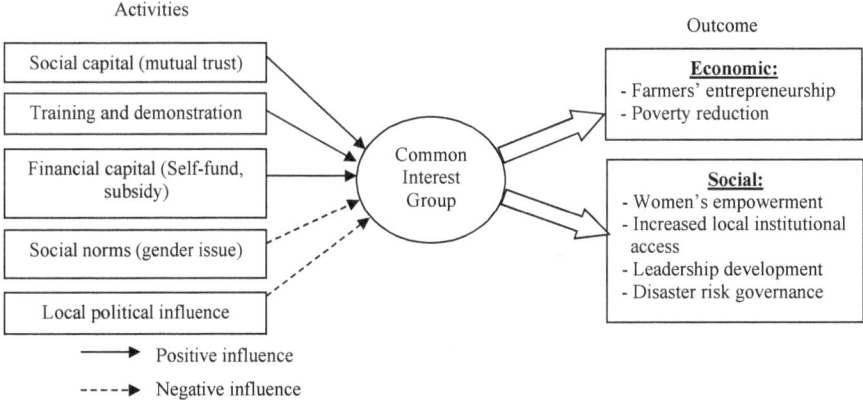

Fig. 6.1 Operational activities and outcome of the CIG approach

social structure (e.g. women are strongly discouraged to work outside, not allowed to operate farm machines and no or limited access to land etc.) women's empowerment is difficult to be strengthen. Moreover, the CIG approach primarily focuses on increasing productivity of agriculture and has limitations in the dynamics of women's empowerment. Finally, the issue of so called 'women's social protection' is also a major concern for their empowerment in the public domain including emergency disaster shelters in the community.

Although the rural women from farming families have been officially recognized as women farmers and increased access to local institutions through the CIG approach, the problems of poverty and gender inequality are complex and firmly route-dependent in the target region (Rana et al. 2020). Therefore, the socio-political transformation beyond the CIG approach is required.

Appendix 6.1: List of Key Informants and Groups Interviewed

See Table 6.2.

Table 6.2 Description of the key informants and methods of interview

Key informants or groups interviewed	Date of interview	Methods of interview
Local agriculture officer	18 August 2019	Interview (face-to-face)
Female CIG (20 members were participated among 30)	19 August 2019	Group discussion (face-to-face)
Union council members (6 members were participated among 13)	19 August 2019	Group discussion (face-to-face)
NGO officer	21 August 2019	Interview (face-to-face)
Local agriculture officer	29 September 2020	Online interview (through Zoom)
Female union council member	28 September 2020	
Male union council member	28 September 2020	
Female CIG leader	01 October 2020	
Male CIG leader	01 October 2020	

Appendix 6.2: Interview with Different Stakeholders Interview with Leader of Male/Female CIGs for the Research on Social Innovation for Women's Empowerment in Disaster Risk Governance in a Haor Region of Bangladesh

This investigation is made for understanding the role of social innovation for women's empowerment in poverty reduction and in disaster risk governance in a haor area of Bangladesh. It will be used only for research purpose and the opinion of participant is anonymous. The personal information will not be in the public. Thank you very much for your cooperation!

Explanation of two concepts:

1. **Social innovation**: Social innovation is defined as a solution to a social problem that is more effective, efficient, sustainable than existing solutions and for which the value created is primarily earned by society rather than private individuals.
2. **Disaster risk governance**: Disaster risk governance means the participation of different stakeholders in decision making and coordinates each other to manage the disaster risks in a given context.

Interview with leader of male/female CIG

Name:…………………………Age:……………..…Sex: Male/Female,
Level of Education:…………………..
Date of interview:…………………………………
Time of interview:…………………………………

Greetings!
Please answer the following questions:
Question 1: When did you form the CIG and how many members are there in your CIG?
Answer:

Question 2: What is socio-economic status of the members of your CIG? (age, literacy level, farm size, annual household income etc.)
Answer:
Age:
Level of education:
Farm size:
Annual household income:

Question 3: What are the activities do you carry out as a CIG? (Multiple answers possible)
Answer:

- Sitting on a monthly meeting

- Discussing contemporary farming issues, problems and solutions
- Participating in training
- Participating in technology demonstration
- Diversifying crop cultivation
- Participating in field day
- Saving money in every month
- Others (specify)

Question 4: What is your role in decision making in CIG as a leader?
Answer:

- Making decision and informing it to the other members
- Discussing with other members to take decision
- Communicating instructions from upper management to the members
- Summarizing the opinions of members and conveying them to upper management
- Others (specify)

Question 5: What changes have you noticed in your CIG members due to the formation of CIG and working in a group? (Multiple answers possible)
Answer:

- Increased access to knowledge and technology
- Increased adoption of improved technology
- Increased the diversity of crop cultivation
- Increased farm production and income
- Increased institutional access at the local level
- Increased access to the local market
- Increased confidence in production and disaster management
- Others (specify)

Question 6: What do you think about organizational partnership in the locality through CIG approach? (Multiple answers possible)
Answer:

- Accessing to different government offices at local level
- Accessing to NGO services (microcredit, health services, legal support etc.)
- Accessing to financial institutions (such as Bank)
- Contacting to input dealers (seed, agro-chemicals etc.)
- Collaborating with other community based organizations (CBOs)
- Others (specify)

Question 7: What do you think about women's empowerment through CIG approach? (Multiple answers possible)
Answer:

- Recognized as individual farmer
- Enhanced leadership skill (speaking and listening to others) due to working in a group
- Participated in income generation activities

- More contributed to household income and nutrition
- Be aware of own problems or rights
- Created a platform to raise their voice
- Accessing to productive assets and household income
- Strengthen the role in household decision-making
- Increased mobility in the community
- Others (specify)

Question 7(1): What are the barriers do you think to women's empowerment?
Answer:

- Lack of awareness of their own capabilities
- Lack of motivation to participate in community development programs
- Cultural norms
- Less access to education and training
- Less access to assets and income
- Lack of decision making power at household/community level
- Restricted mobility in the community
- No/less access to local market
- Others (specify)

Question 8: Do you have any interaction with other CIGs (male or female)?
Answer:

- Male (Yes, No)
- Female (Yes, No)

If 'Yes' how? If 'No' why?

Question 9: Do you face any problems while working with CIG? (Multiple answers possible)
Answer:

- Conflict among group members (for reasons like position in the group, unequal sharing of advantages etc.)
- Not all members are equally interested in the same subject/technology
- Political influence of local government (union council) elections
- Less cooperation from agriculture office
- Others (specify)

Question 10: How can you solve these problems (to what extent)?
Answer:

Question 11: Discuss the role of women in disaster risk governance in the community (Multiple answers possible)
Answer:

- They can survive and acquire coping skills in emergency situations including food preservation
- They preserve cooking materials for disaster period

- They have family and community roles that make them important risk communicators
- Their social networks provide information about members of their community who may be need assistance, or who can help in times of crisis
- They play a leadership role in local networks and organizations
- They have access to in-depth information that is often not easily accessible to outsiders
- They can contribute to the empowerment of individuals and families
- They can play an important role in monitoring development programs
- Others (specify)

Question 12: Do you think the CIG approach has any impact on the role of women in disaster risk governance in the community?
Answer:

- Yes
- No

If 'Yes' how?...........................

Question 13: What are your expectations through CIG approach for women's empowerment in DRG? (Multiple answers possible)
Answer:

- More training on............
- Credit support from formal financial institutions
- Easy access to local market
- Bargaining power at household as well as community level
- Access to other social services (e.g. health and sanitation)
- Awareness campaign
- Others (specify)

Interview with local agriculture officer for the research on Social Innovation for Women's Empowerment in Disaster Risk Governance in a Haor Region of Bangladesh

This investigation is made for understanding the role of social innovation for women's empowerment in poverty reduction and in disaster risk governance in a haor area of Bangladesh. It will be used only for research purpose and the opinion of participant is anonymous. The personal information will not be in the public. Thank you very much for your cooperation!

Explanation of two concepts:

1. **Social innovation**: Social innovation is defined as a solution to a social problem that is more effective, efficient, sustainable than existing solutions and for which the value created is primarily earned by society rather than private individuals.

2. **Disaster risk governance**: Disaster risk governance means the participation of different stakeholders in decision making and coordinates each other to manage the disaster risks in a given context.

Interview with local agriculture officer

Name:…………………………..………..Age…………...Sex: Male/Female,
Date of interview:…………………………..
Time of interview:…………………………

Greetings!
Please answer the following questions:

Question 1: How CIGs are formed? What are the criteria for the farmers to be a member of CIG?
Answer:

Question2: Why male and female CIGs are separate?
Answer:

Question 3: Do you have any female Sub-Assistant Agriculture Officer (SAAO) in your office to work with female CIGs?
Answer:

- Yes
- No

If 'No' why?…..

Question 4: Have you noticed any change in CIG members due to working in a group? (Multiple answers possible)
Answer:

- Increased access to farming knowledge and improved technology
- Increased adoption of improved technology
- Increased the diversity of crop cultivation
- Increased farm production and income and reduce poverty
- Increased institutional access at the local level
- Increased access to the local market
- Others (specify)

Question 5: What do you think about women's empowerment through CIG approach? (Multiple answers possible)
Answer:

- Recognized as individual farmer
- Enhanced leadership skill (speaking and listening to others) due to working in a group
- Participated in income generation activities
- More contributed to household income and nutrition

- Be aware of own problems or rights
- Created a platform to raise their voice
- Accessing to productive assets and household income
- Strengthen the role in household decision-making
- Increased mobility in the community
- Others (specify)

Question 6: Do you think the collaboration at organizational level has been increased in the locality through CIG approach?
Answer:

- Very much
- A little
- Not at all
- I cannot say

Question 7: Do you think the CIG approach has any impact on disaster risk governance in the community?
Answer:

- Yes
- No

If 'Yes' how? If 'No' why?.....................

Question 8: What are the challenges or limitations of CIG approach?
Answer:

Question 9: What are the strategies to overcome these challenges or limitations?
Answer:

Interview with union council member (female) for the research on Social Innovation for Women's Empowerment in Disaster Risk Governance in a Haor Region of Bangladesh

This investigation is made for understanding the role of social innovation for women's empowerment in poverty reduction and in disaster risk governance in a haor area of Bangladesh. It will be used only for research purpose and the opinion of participant is anonymous. The personal information will not be in the public. Thank you very much for your cooperation!

Explanation of two concepts:

1. **Social innovation**: Social innovation is defined as a solution to a social problem that is more effective, efficient, sustainable than existing solutions and for which the value created is primarily earned by society rather than private individuals.
2. **Disaster risk governance**: Disaster risk governance means the participation of different stakeholders in decision making and coordinates each other to manage the disaster risks in a given context.

Interview with female union council member

Name:…………………………………Age:……………..…………Sex: Male/Female,
Level of Education…..……..
Date of interview:……………………………………
Time of interview:…………………………………

Greetings!
Please answer the following questions:

Question 1: In which year were you elected as a member of the union council? How many competitors were you in the election? What is your strong point to win by your evaluation?
Answer:

Question 2: What are the existing practices for women's empowerment and poverty reduction in your region? (Multiple answers possible)
Answer:

- Microcredit support for income generation activities
- Formation of common interest group (CIG)
- Access to local market to sell their products directly
- Maternity allowance from local government
- Allowance for widow from local government
- Legal assistance to women against violence
- Others (specify)

Question 3: Do you think that these practices have any effect on the empowerment of women and poverty reduction in the society?
Answer:

- Yes
- No

If 'Yes' what kind of effects?

- Access to productive assets and household income
- Participation in income generation activities
- More contribution to household income and nutrition
- Be aware of own problems or rights
- Strengthen their role in household decision-making
- Increase mobility in the community
- Increase access to health and sanitation services
- Others (specify)

Question 4: What are the common disaster risk management practices in your locality?
Answer:

- Construction of flood control dam (concrete and or earthen)

- Establishment of community tube well for safe drinking water
- Vulnerable group feeding (VGF)
- Relief distribution
- Disaster shelter management
- Spreading early warning message from union council
- Others (specify)

Question 5: What is the level of organizational collaboration (GOs, NGOs, CBOs etc.) in the activities related to women's empowerment and disaster risk governance?
Answer:

- Very much
- A little
- Not at all
- I cannot say

Question 6: Explain the role of women in disaster risk governance at the community level (Multiple answers possible)
Answer:

- They can survive and acquire coping skills in emergency situations including food preservation
- They preserve cooking materials for disaster period
- They have family and community roles that make them important risk communicators
- Their social networks provide information about members of their community who may be need assistance, or who can help in times of crisis
- They play a leadership role in local networks and organizations
- They have access to in-depth information that is often not easily accessible to outsiders
- They can contribute to the empowerment of individuals and families
- They can play an important role in monitoring development programs
- Others (specify)

Question 7: How do you play your role as a council member in community development activities?
Answer:

Question 8: Can you express your opinion freely in the union council meeting?
Answer:

- Yes
- No

If "No" why....

Question 9: How much your opinion is accepted in the union council meeting?
Answer: ....%

If your opinion is rejected, what is your general reaction?

Answer:

- Getting angry or depressed
- Trying not to worry
- Challenging again

Question 10: Give your opinion on how to change or improve the existing situation of women's empowerment and disaster risk governance in your region?
Answer:

Interview with union council member (male) for the research on Social Innovation for Women's Empowerment in Disaster Risk Governance in a Haor Region of Bangladesh

This investigation is made for understanding the role of social innovation for women's empowerment in poverty reduction and in disaster risk governance in a haor area of Bangladesh. It will be used only for research purpose and the opinion of participant is anonymous. The personal information will not be in the public. Thank you very much for your cooperation!

Explanation of two concepts:

1. **Social innovation**: Social innovation is defined as a solution to a social problem that is more effective, efficient, sustainable than existing solutions and for which the value created is primarily earned by society rather than private individuals.
2. **Disaster risk governance**: Disaster risk governance means the participation of different stakeholders in decision making and coordinates each other to manage the disaster risks in a given context.

Interview with male union council member

**Name:……………………………………Age:……………………………Sex: Male/Female,
Level of Education…………..
Date of interview:……………………………………
Time of interview:……………………………………**

Greetings!
Please answer the following questions:

Question 1: In which year were you elected as a member of the union council? How many competitors were you in the election? What is your strong point to win by your evaluation?
Answer:

Question 2: What are the existing practices for women's empowerment and poverty reduction in your region? (Multiple answers possible)
Answer:

- Microcredit support for income generation activities
- Formation of common interest group (CIG)

- Access to local market to sell their products directly
- Maternity allowance from local government
- Allowance for widow from local government
- Legal assistance to women against violence
- Others (specify)

Question 3: Do you think that these practices have any effect on the empowerment of women and poverty reduction in the society?
Answer:

- Yes
- No

If 'Yes' what kind of effects?

- Access to productive assets and household income
- Participation in income generation activities
- More contribution to household income and nutrition
- Be aware of own problems or rights
- Strengthen their role in household decision-making
- Increase mobility in the community
- Increase access to health and sanitation services
- Others (specify)

Question 4: What are the common disaster risk management practices in your locality?
Answer:

- Construction of flood control dam (concrete and or earthen)
- Establishment of community tube well for safe drinking water
- Vulnerable group feeding (VGF)
- Relief distribution
- Disaster shelter management
- Spreading early warning message from union council
- Others (specify)

Question 5: What is the level of organizational collaboration (GOs, NGOs, CBOs etc.) in the activities related to women's empowerment and disaster risk governance?
Answer:

- Very much
- A little
- Not at all
- I cannot say

Question 6: Explain the role of women in disaster risk governance at the community level (Multiple answers possible)
Answer:

- They can survive and acquire coping skills in emergency situations including food preservation
- They preserve cooking materials for disaster period
- They have family and community roles that make them important risk communicators
- Their social networks provide information about members of their community who may be need assistance, or who can help in times of crisis
- They play a leadership role in local networks and organizations
- They have access to in-depth information that is often not easily accessible to outsiders
- They can contribute to the empowerment of individuals and families
- They can play an important role in monitoring development programs
- Others (specify)

Question 7: How do you play your role as a council member in community development activities?
Answer:

Question 8: How much do you value the participation of women members in union council activities?
Answer:

- Extremely important
- Very important
- Important
- Somewhat important
- Not important

Question 9: How can you evaluate the performance of women council members in community development activities?
Answer:

- Very good
- Good
- Average
- Poor
- I cannot say

Why?.....................

Question 10: Give your opinion on how to change or improve the existing situation of women's empowerment and disaster risk governance in your region?
Answer:

References

Moulaert F, MacCallum D (2019) Advanced introduction to social innovation. Edward Elgar Publishing, Cheltenham, UK and Northampton, MA, USA

Rana S, Kiminami L, Furuzawa S (2020) Analysis on the factors affecting farmers' performance in disaster risk management at community level: focusing on a haor locality in Bangladesh. Asia-Pacific J Reg Sci 4(3):737–757

Chapter 7
Performance of Farm Households in Disaster Risk Management at Community Level

Abstract The purpose of this chapter is to clarify the factors affecting famers' performance in disaster risk management (DRM) at the community level. We introduced structural equation modeling (SEM) based on the results of the authors' designated questionnaire survey (face-to-face interview) of farm households in the target region, to verify our hypothesis "Farmers' performance in disaster risk management at community level is determined by their socio-economic attributes, social capital, and access to local institutions (H3)". The results clarified that there are multi-causal relationships among the factors affecting farmers' performance in DRM, in which, the factors of gender, poverty, social capital, and farmers' attitudes toward technology and training are important in the region. Therefore, policy implications suggest to recognize the causes of poverty strongly route-dependent in rural areas including the existence of gender discrimination and low trust in government officials.

Keywords Farm households · DRM · Community level · SEM · Sustainability

7.1 Questionnaire Survey

We investigated the farm households of Itna union (the lowest unit of local government) in the Kishoreganj district of Bangladesh. We interviewed (face-to-face) ramdomly selected 150 farm households out of 3,028 using a structured questionnaire (see Appendix 7.1) during August 2019. One adult member either male or female was interviewed from each household. Among the respondents, 49 were women and 101 were male, and 51 were household heads among the male. A four-point scale (0–3; low to high) was constructed to assign scores based on the performance of the respondents in each aspect of livelihood for every question regarding DRM and to understand the respondents' institutional access as well as the contact with selected information sources for the technology related to livelihood development. A Five-Point Likert Scale was used to measure their attitudes toward agricultural technologies.

7.2 Structural Equation Modeling (SEM)

Since our hypothesis is to explain the impacts of different factors on farmers' performance in DRM at community level, which is a complex phenomenon, we introduced structural equation modeling (SEM) that can represent the relationship among variables with path coefficients (Hox and Bechger 1999) for hypothesis verification. We used the R software version 3.5.2 (free) for conducting SEM analysis.

We set the following analytical framework (Fig. 7.1) and variables settings (Table 7.1) considering the socio-economic situation of the farm households in the context of rural Bangladesh based on the literature review to verify the (H3) "Farmers' performance in disaster risk management at community level is determined by their socio-economic attributes, social capital, and access to local institutions". Different dimensions of livelihood, such as crop production, livestock and poultry, housing and shelter, health and sanitation, and livelihoods and food security are considered in the research for assessing famer's performance in DRM (Islam 2005; Rana et al. 2010, 2019). Both the explanation of the variables and descriptive statistics of the variables used in the SEM analysis are shown in Table 7.1 and Table 7.2, respectively.

7.3 Expected Sings of the Variables in SEM Analysis

Table 7.3 expresses the expected signs of the variables in SEM analysis. Since income is one of the important indicators related to overall development of human life, annual household income is expected to have a positive impact on the performance of disaster risk management. As for farm size, it is also expected to have a positive impact on DRM as it is linked with agricultural production and household income. As for age, the results of survey revealed that older age groups have a higher tendency to

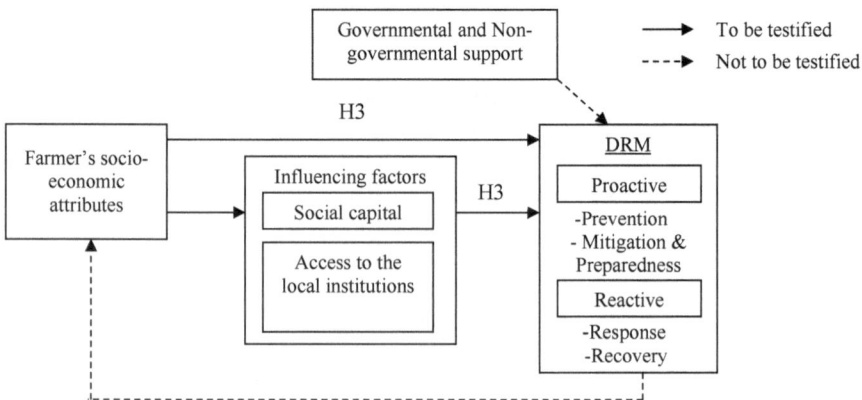

Fig. 7.1 Analytical framework for SEM analysis

Table 7.1 The explanation of the variables used in the SEM analysis

Variables	Explanation	References for variable setting
Socio-economic attributes		
Age [AGE]	The age of a respondent (years)	Cabinet Office (2017), Rana et al. (2019), Cvetkovic et al. (2018), Akter and Mallick (2013)
Education [EDU]	Formal year of schooling. But 0.5 score was assigned who can sign only	
Sex [SEX]	Female (0) and male (1)	
Farm size [FS]	The total area of farm on which he/she continued his farming operations (Ha)	
Annual household income [AHI]	The annual gross income of members of the household (thousand Taka)	
Social capital		
SC	Score of Principal Component 1 (level of social capital; low-high) (see Appendix 7.2)	Rana et al. (2019), Islam and Walkerden (2015, 2017)
Access to the local institutions		
Extension media contact [EMC]	The access of the respondents to local institutions and the contact with different sources for receiving technological information	Rana et al. (2019)
Attitudes toward modern agricultural technologies [AMT]	The statements (level of agreement) related to agricultural technologies (in the context of study area) based on a Five-Point Likert Scale	
Training [TRAIN]	The number of days participated in the training programs	Kazal et al. (2010)
Farmer's performance in disaster risk management		
Prevention [PREVN]	A four-point scale for five questions in each stage of disaster risk management was assigned based on their performance in DRM	Islam (2005), Rana et al. (2010, 2019)
Mitigation & Preparedness [MITPD]		
Response [RESPO]		
Recovery [RECOV]		

prepare for disasters than younger age groups in Japan (Cabinet Office 2017). Unfortunately, we could not find literatures on the effect of general education on disaster risk management. According to Kazal et al. (2010), skill training related to livelihood development plays a vital role in facilitating the income generation activities in *haor* areas of Bangladesh. Since positive attitudes toward modern agricultural technologies are supposed to have positive impacts on productivity of agriculture and

Table 7.2 Descriptive statistics of the variables used in the SEM

Variables	Layer	No of obs.	Mean	SD	Min.	Max.
Socio-economic attributes						
Age	(1)	150	41.55	10.598	24	70
Education	(1)	150	4.52	3.889	0	14
Sex	(1)	150	0.67	0.470	0	1
Farm size	(1)	150	2.43	2.776	0.49	13.85
Annual household income (thousand Taka)	(1)	150	283.33	280.459	60	1400
Social capital						
SC	(2)	150	1.88	1.071	0	3.345
Access to the local institutions						
Extension media contact	(2)	150	10.66	3.516	4	27
Attitudes toward modern agricultural technologies	(2)	150	25.13	1.995	18	30
Training	(2)	150	1.18	1.447	0	4
Farmer's performance in disaster risk management						
Prevention	(3)	150	7.98	1.762	5	13
Mitigation & preparedness	(3)	150	7.20	1.676	5	12
Response	(3)	150	7.22	1.760	4	12
Recovery	(3)	150	6.78	1.673	4	12

Note Layer indicates the position of the variables in the analytical framework (Fig. 7.1)

farm income, it is expected to have a positive impact on the farmer's performance in disaster risk management as well. Finally, social capital is also expected to have a positive impact on farmer's performance in DRM based on the literature review.

Table 7.3 Expected sings of the variables in SEM

Variables	Expected sign
Age	Positive (+)
Education	Positive (+), Negative (−)
Sex	Positive (+), Negative (−)
Farm size	Positive (+)
Annual household income	Positive (+)
Social capital	Positive (+)
Access to the local institutions	Positive (+)

7.4 Results of the SEM Analysis

We introduced the structural equation model (SEM) analysis to the results obtained from the questionnaire survey to explain the relations among the variables. Figure 7.2, Tables 7.4 and 7.5 show the path diagram and path coefficient respectively obtained from the SEM analytical results. The goodness of fit indicators is RMSEA = 0.245 and CFI = 0.942, which indicates a good result. According to the analytical results, the relations of variables with disaster risk management are explained as follows.

First, among the attributes of farmers, annual household income had a strong positive impact on the prevention of DRM. Age had aweak negative impact on the response of DRM, which is different from the result of the survey in Japan. However, age square had a positive impact on the response indicated that both famers' physical capability and experience from disasters can affect their performance in DRM. Education had a positive impact on the performance in the stage of prevention, and annual household income had a positive impact on all stages of DRM. However, farm size had no direct impact on famer's performance in DRM but was strongly correlated with annual household income and it was also likely to determine the status of the respondent in the community.

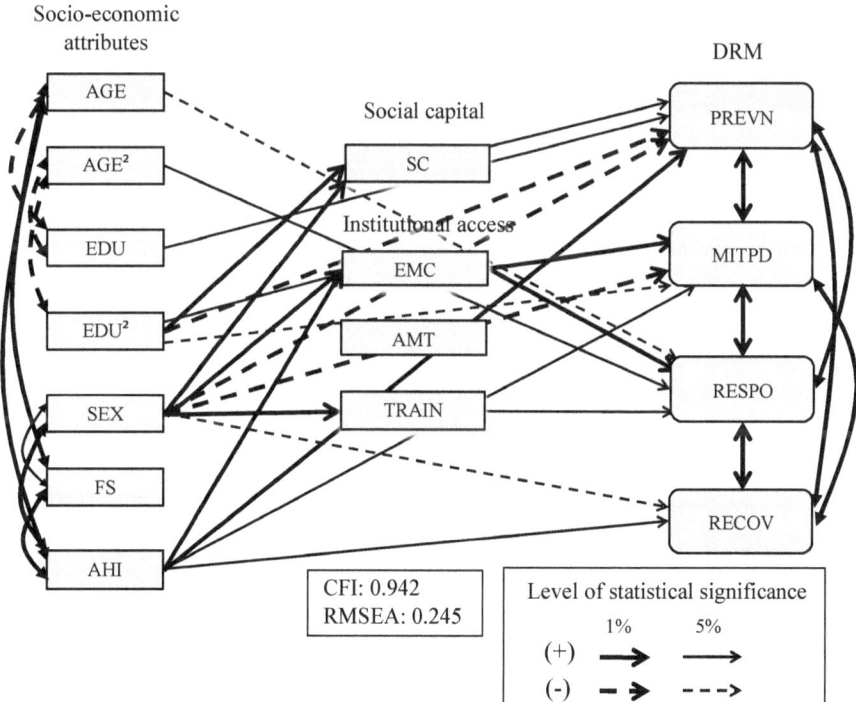

Fig. 7.2 Path diagram showing the relations among variables

Table 7.4 Results of SEM analysis

| | | Disaster risk management (layer 3) | | | | | | | | Layer 2 | | | | | | | |
| | | vezePREVN | | MITPD | | RESPO | | RECOV | | SC | | EMC | | AMT | | TRAIN | |
		Stand. coef.	P > |z|	Stand. coef.	P > |z|	Stand. coef.	P > |z|	Stand. coef.	P > |z|	Stand. Coef.	P > |z|	Stand. Coef.	P > |z|	Stand. coef.	P > |z|	Stand. coef.	P > |z|
Socio-economic attributes (layer 1)	AGE	-0.231	0.520	-0.256	0.508	-0.992**	0.018	0.276	0.485	-0.188	0.713	-0.069	0.870	-0.165	0.776	0.194	0.731
	AGES	0.305	0.379	0.273	0.463	0.934**	0.021	-0.139	0.715	-0.134	0.786	0.050	0.902	0.106	0.850	-0.301	0.581
	EDU	0.361**	0.034	0.250	0.172	-0.027	0.892	0.257	0.171	-0.462*	0.053	-0.079	0.689	0.072	0.792	0.069	0.793
	EDUS	-0.519***	0.002	-0.379**	0.033	-0.096	0.623	-0.295	0.106	0.713***	0.002	0.391**	0.036	0.026	0.918	0.076	0.760
	SEX	-0.246***	0.000	-0.181***	0.002	0.067	0.289	-0.122**	0.041	0.205***	0.004	0.184***	0.002	-0.026	0.746	0.213***	0.007
	FS	-0.420*	0.054	0.057	0.807	0.081	0.752	0.125	0.604	0.306	0.316	-0.249	0.323	0.556	0.109	0.224	0.506
	AHI	1.149***	0.000	0.608**	0.012	0.470*	0.012	0.546**	0.027	0.079	0.797	0.813***	0.001	-0.164	0.641	0.119	0.728
Social capital	SC	0.122**	0.033	0.012	0.846	-0.063	0.349	0.064	0.315	–	–	–	–	–	–	–	–
Institutional access	EMC	0.055	0.430	0.207***	0.005	0.234***	0.004	0.068	0.377	–	–	–	–	–	–	–	–
	AMT	-0.018	0.719	-0.029	0.597	-0.012	0.842	0.061	0.270	–	–	–	–	–	–	–	–
	TRAIN	0.059	0.258	0.070	0.212	0.132**	0.030	-0.004	0.942	–	–	–	–	–	–	–	–
Goodness of fit	RMSEA	0.245															
	CFI	0.942															

Note * (10%), ** (5%) and *** (1%) indicate the statistical significance

Details of the variable are as follows

Age: years; Education: score 1 for each year of schooling, for who can sign only (0.5); Sex: female (0), male (1); Farm size: area of farming operations (Acre); Annual household income: household income in thousand Taka

Table 7.5 Covariance

		Layer	Coef.	P > |z|
AGE (1)	AGES	(1)	0.983***	0.000
	EDU	(1)	−0.473***	0.000
	EDUS	(1)	−0.375***	0.000
	SEX	(1)	0.298***	0.000
	FS	(1)	0.328***	0.000
	AHI	(1)	0.315***	0.000
AGES (1)	EDU	(1)	−0.429***	0.000
	EDUS	(1)	−0.329***	0.000
	SEX	(1)	0.300***	0.000
	FS	(1)	0.314***	0.000
	AHI	(1)	0.304***	0.000
EDU (1)	EDUS	(1)	0.945***	0.000
	SEX	(1)	0.019	0.817
	FS	(1)	−0.005	0.950
	AHI	(1)	0.060	0.459
EDUS (1)	SEX	(1)	0.024	0.768
	FS	(1)	0.019	0.812
	AHI	(1)	0.086	0.289
SEX (1)	FS	(1)	0.198**	0.017
	AHI	(1)	0.222***	0.008
FS (1)	AHI	(1)	0.967***	0.000
PREVN (3)	MITPD	(3)	0.149***	0.000
	RESPON	(3)	0.123***	0.000
	RECOV	(3)	0.105***	0.000
MITPD (3)	RESPO	(3)	0.206***	0.000
	RECOV	(3)	0.161***	0.000
RESPO (3)	RECOV	(3)	0.215***	0.000

Note * (10%), ** (5%) and *** (1%) indicate the statistical significance

Secondly, sex had a strong negative impact on the stages of prevention, and mitigation and preparedness of DRM indicated that women contributed more to disaster risk management at community level, although they received less education and training, had a less access to assets and a lower level of social capital than their male counterpart.

Thirdly, social capital had a positive impact on farmer's performance in disaster risk management. Among the respondents, 88% reported that they are trying to be helpful to others in their community. More than three fourths (78%) of the respondents trusted one another in their community (Table 7.6). The individuals in a community

Table 7.6 Response concerning the social capital

Question	Response (N = 150)	
Q1. Would you say that most of the time, people try to be helpful or that they are mostly looking out for themselves?	Try to be helpful 132 (88%)	Look out for themselves 18 (12%)
Q2. Generally speaking, would you say that most people can be trusted or you had better to be careful in dealing with people?	Can be trusted 117 (78%)	Had better to be careful 33 (22%)
Q3. Do you think that you can trust the community leader?	Yes 78 (52%)	No 72 (48%)
Q4. Do you have the interaction with local government officials through community development activities?	Yes 61 (40.7%)	No 89 (59.3%)
Q5. Do you think that you can trust the local government officials?	Yes 58 (38.7%)	No 92 (61.3%)

with a strong mutual trust have high motivation to take some sort of preventive actions prior to disasters (Dynes 2006; Joerin et al. 2012). On the other hand, their trust to the local government officials was low. Similar to these findings, Islam and Walkerden (2017) noted that the three key weaknesses of DRM policies, such as top-down policy formulation, lack of coordination and corruption had hindered the development of social capital at community level and had negative impacts on the recovery from disasters in rural Bangladesh.

Fourthly, farmers' access to the local institutions had positive impacts on their performance in the stages of mitigation and preparedness, and response of DRM, respectively. It indicated that, farmers' positive attitudes toward new agricultural technologies and training received in various aspects of their livelihood as a direct or indirect outcome of the access to the local institutions that contributed to the increase in farm production and income, had positive impacts not only on a better performance in DRM but also other dimensions of their lives.

Appendix 7.1: Interview Guide of Farm Households Questionnaire for the Research on Disaster Risk Management (DRM) in a Haor Area of Bangladesh: Focusing on the Farm Household Level

This investigation is made for understanding the disaster risk management of farm households in a haor area of Bangladesh. It will be used only for paper and the opinion of participant is anonymous. The personal information will not be in the public. Thank you very much for your cooperation!

Please furnish the information as mentioned in the following items

1. Age ( ) years
2. Gender (male/female)
3. Education ( )
4. Member of the household

2019		2018		2017	

5. Farm size ( )
6. 6. Annual household income, BDT

2019		2018		2017	

7. Capacity of stock piling of food materials: Yes No [If Yes.....................Months]
8. Social capital

A. Bonding social capital

Question	Response	
Q1. Would you say that most of time, people try to be helpful, or that they are mostly just looking out of themselves?	Try to be helpful	Look out for themselves
Q2. Generally speaking, would you say that most people can be trusted or you had better to be careful in dealing with people?	Can be trusted	Had better to be careful
Q3. Do you think that you can trust the community leader?	Yes	No

B. Bridging social capital: Please indicate your membership to different community-based organizations (CBOs) including NGOs

Sl. no.	Organization	Duration (year)	Nature of membership	
			Ordinary member	Executive position
1.				
2.				
3.				

C. Linking social capital

Question	Response	
Q1. Do you have the Interaction with local government officials through community development activities?	Yes	No
Q2. Do you think that you can trust the local government officials?	Yes	No

D. Locale orientation

For how many years have you been living in this locality? ( ) years

9. Technological attitude of the farmers

A. Extension media contact: Please indicate your frequency of communication with the following extension media

Sl. no.	Name of extension media	Extent of contact			
		Frequently	Occasionally	Rarely	Not at all
A. Individual contact					
1.	Upazila level officers (UAO/AEO/UFO/ULO/VS) *(per 3 months)*				
2.	Sub Assistant Agriculture Officer (SAAO, formerly known as BS) *(per 3 months)*				
3.	NG4O workers/ field agents *(per 3 months)*				
4.	Health workers and other public sector field agents *(per 3 months)*				
5.	Input dealer (fertilizer, seed, feed, pesticide etc.) *(per 3 months)*				
B. Group contact					
6.	Group discussion with change agents *(per year)*				
7.	Participation in training and method demonstration *(per year)*				
8.	Participation in result demonstration meetings *(per year)*				
C. Mass contact					
9.	Listening to development programmes over Radio *(per month)*				
10.	Watching TV for development programmes *(per month)*				
11.	Reading related books /magazines /leaflets etc. *(per six months)*				

Note Score 3 = Frequently ($\geq$6 times), 2 = Occasionally (3–5 times), 1 = Rarely (1–2 times) and 0 = Not at all (0)

We sum up the score of all items to make the variable.

B. Attitude towards modern agricultural technologies

Statements	Extent of response				
	SA	A	N	D	SD
1. I like USG because it reduces total amount of urea required for rice cultivation					
2. I don't like USG because it requires more labour					
3. I like hybrid varieties because its production is more than local varieties					
4. High price of hybrid rice seed is a problem for wide scale cultivation					
5. Artificial insemination is beneficial for more milk production					
6. Chemical fertilizer increases the production of rice					
7. Use of pesticides is harmful for environment					
8. Excess use of chemical fertilizers is harmful for long term soil fertility					

Note Score 4 = Strongly agree (SA), 3 = Agree (A), 2 = Neutral (N), 1 = Disagree (D), 0 = Strongly disagree (SD) has been applied for the positive statements. But opposite order of scoring, such as 4 = Strongly disagree (SD), 3 = Disagree (D), 2 = Neutral (N), 1 = Agree (A), 0 = Strongly agree (SA) has been applied for two negative statements (Nos. 2 and 4). Then we sum up the score of all statements to make the variable

10. Disaster risk management (DRM) (Open type questions)

What do you usually do in the following situations?

Situation	Score*			
(A) Prevention	Score (0–3)			
To avoid loss of standing crops from flood and other disasters (precaution, cropping pattern etc.)	3	2	1	0
To avoid damage of livestock and poultry birds in advance	3	2	1	0
To protect house and homestead assets from disaster	3	2	1	0
To prevent health related problems during and after disaster	3	2	1	0
To secure food materials during and after disaster	3	2	1	0
(B) Mitigation and preparedness				
To mitigate crop damage by disaster	3	2	1	0

(continued)

(continued)

Situation	Score*			
To preserve feed for cattle and poultry birds in advance	3	2	1	0
To adjust situation when house is inundated/destroyed	3	2	1	0
To have preparation for different diseases	3	2	1	0
To procure major food items	3	2	1	0
(C) Response				
To save standing crops	3	2	1	0
To manage shelter and feed for cattle and poultry	3	2	1	0
To evacuate family members and assets	3	2	1	0
To secure pure drinking water	3	2	1	0
Search for job to earn money	3	2	1	0
(D) Recovery & development				
To overcome damage of standing crops	3	2	1	0
To overcome damage of livestock	3	2	1	0
To restore damaged houses immediately after disaster	3	2	1	0
To make advance plan before disaster for food security	3	2	1	0
To make advance plan for securing pure drinking water after disasters	3	2	1	0

[a]*Note* Score 0 = Do nothing, 1 = To a very limited extent, 2 = Some activity but significant scope for improvement, 3 = Satisfactory and effective measures in place

We sum up the score of five questions in each stage of DRM to make the variable.

11.   Coping strategies for household food security of farmers

Sl. no.	Coping strategy	Regularly	Occasionally	Rarely	Not at all
1	Rely on less preferred and less expensive food items				
2	Borrow food from neighbor/relatives/friends				
3	Purchase food on credit				
4	Reduce portion sizes at mealtime				
5	Reduce number of meals in a day				
6	Rely on help from friend or relatives				
7	Out migration of household members				
8	Borrow money from NGOs/GOs				
9	Sell labour in advance				
10	Spend money from deposit				
11	Mortgage land and ornaments				
12	Sell land and other assets (tools, seeds, livestock etc.)				

12.   ICT based initiatives

A.   Information about ICT based initiatives

Question	Response	
Q1. Do you use mobile phone?	Yes	No
Q2. Do you have the Smartphone?	Yes	No
Q3. Any member of your household has the Smartphone?	Yes	No
Q4. Do you have the internet access?	Yes	No
Q5. Do you use any apps regarding agricultural or other purposes?	Yes	No
Q6. Do you receive any early warning message about disaster?	Yes	No
Q7. Do you search agricultural information by using ICT tools?	Yes	No

B.   Frequency of the visit of AICC/UDC

- Everyday
- Once a week
- Once a month
- 2–4 times per month
- Rarely
- No response

C.   What type of information/service you received from AICC/UDC?

- Crop cultivation related information
- Livestock rearing related information
- Fisheries related information
- Market and price information
- Agriculture related leaflet or booklet collection
- Government forms

13.   Training

Did you receive any training from any government agency or NGO? Yes/No
   If yes, please give the following information

Sl. no.	Name of the training course	Organization	Duration (days)	Fee (Tk.)
1.				
2.				
3.				

14.   Sustainable development issues of agriculture

Aspects of sustainability issues	Yes	No
(a) Environmental aspect		
1. Do you have the facilities for safe water for domestic use?		
2. Do you maintain effective resource/input management in farming?		
3. Do you consider the soil conservation practices in your farming?		
4. Do you maintain the biodiversity (plant and tree diversity)?		
5. Do you able to recycle of homestead garbage?		
(b) Economic aspect		
1. Do you think that your farming income is increased by time?		
2. Do you have the access to market information?		
3. Do you have the access to credit?		
4. Do you have the off farm business opportunities?		
(c) Social aspect		
1. Do you have the access to health services?		
2. Do you think that schooling of children is getting better?		
3. Do you have the access to receive training on different subjects?		
4. Do you think that your food security status is improved?		
5. Do you have the access to the assets including land?		
6. Do you have the access to sanitation facilities?		

15. Do you think that the quality of life has been improved during last ten years?

SA	A	N	D	SD

SA: Strongly agree, A: Agree, N: Neutral, D: Disagree, SD: Strongly disagree

16. To your evaluation describe the role of women regarding disaster risk management.
17. Please give your opinion how better disaster risk management could be ensured in your locality.

Appendix 7.2: Principal Component Analysis (PCA) of Social Capital Variable

See Table 7.7.

Table 7.7 Results of the principal component analysis on the social capital

	Score of principal component 1
1. Would you say that most of the time, people try to be helpful or that they are mostly looking out of themselves?	0.549
2. Generally speaking, would you say that most people can be trusted or you had better to be careful in dealing with people?	0.600
3. Do you think that you can trust the community leader?	0.541
4. Do you have the interaction with local government officials through community development activities?	0.824
5. Do you think that you can trust the local government officials?	0.831
Ratio of contribution	46.47%
Interpretation of component	Level of social capital (low-high)

References

Akter S, Mallick B (2013) The poverty–vulnerability–resilience nexus: evidence from Bangladesh. Ecol Econ 96:114–124

Cabinet Office (2017) White paper disaster management in Japan 2017, Japan

Cvetkovic VM, Roder G, Ocal A, Tarolli P, Dragicevic S (2018) The role of gender in preparedness and response behaviors towards flood risk in Serbia. Int J Environ Res Public Health 15:1–21

Dynes R (2006) Social capital: dealing with community emergencies. Homeland Secur Aff 2(2):1–26

Hox JJ, Bechger TM (1999) An introduction to structural equation modeling. Family Sci Rev 11:354–373

Islam MS (2005) Farmer's ability to cope with the flood in a selected area of Jamalpur district. MS thesis, Department of Agricultural Extension Education, Bangladesh Agricultural University, Mymensingh

Islam R, Walkerden G (2015) How do links between households and NGOs promote disaster resilience and recovery?: A case study of linking social networks on the Bangladeshi coast. Nat Hazards 78(3):1707–1727. https://doi.org/10.1007/s11069-015-1797-4

Islam R, Walkerden G (2017) Social networks and challenges in government disaster policies: a case study from Bangladesh. Int J Disast Risk Reduct 22:325–334

Joerin J, Shaw R, Takeuchi Y, Krishnamurthy R (2012) Assessing community resilience to climate-related disasters in Chennai, India. Int J Disast Risk Reduct 1(1):44–54

Kazal MMH, Villinueva CC, Hossain MZ, Das TK (2010) Food security strategies of the people living in haor areas: status and prospects. National Food Policy Capacity Strengthening Program (NFPCSP), FAO, Dhaka

Rana S, Islam MN, Rahman MH (2010) Disaster management ability of farmers in a selected haor area of Kishoreganj district. J Environ Sci Nat Resourc 3(1):157–161

Rana S, Kiminami L, Furuzawa S (2019) Disaster risk management (DRM) in a haor area of Bangladesh: focusing on the household level. In: The 56th annual meeting of the Japan Section of Regional Science Association International, 13–15 September 2019, Kurume University, Kurume City, Japan

Chapter 8
General Conclusions and Policy Implications

Abstract We obtained the following conclusions from the empirical analysis in the previous chapters. First, the results from TEM analysis based on three male and three female cases clarified that socio-cultural and political changes in the relationships between men and women, as well as among the same sexuality are necessary. Secondly, the CIG (common interest group) approach created a platform for the rural farming communities to get access to the local institutions through the accumulation of social capital. In addition, the rural women of farming households are officially recognized as female farmers through the CIG approach although the entrepreneurial development of women farmers is still limited in the region. Thirdly, there are multi causal relationships among the factors affecting farmers' performance in DRM, in which, the factors of gender, poverty, social capital and farmers' attitudes toward technology and training are important in the *haor* area of Bangladesh. Therefore, our policy implication suggests that the government should pay more attention to the entrepreneurship and social innovation for socio-cultural and political transformation through changing the collective cognition of the society along with economic growth.

Keywords Sustainable regional development · Entrepreneurship · Social innovation · *Haor* region · Bangladesh

Sustainable development is a continuous cyclical process, and in a particular region, it depends on many things to do well, and coordinate each other in the context responsive ways. In order to achieve sustainable development, the following aspects such as economic efficiency, socializatio, and environmental protection must be considered broadly at the same time. The main purpose of the study is to assess the role of entrepreneurship and social innovation for socio-cultural changes, and farmers' performance in disaster risk management at the community level towards sustainable regional development in a less favored *haor* region of Bangladesh. The following conclusions are obtained from the empirical investigations.

First, based on the above-mentioned TEM analytical results, our hypothesis 1 "Social and cultural changes through the development of local entrepreneurship is necessary for the sustainable regional development (H1)" has been verified. The TEM analytical results also clarified that social and cultural changes in the relationships are necessary not only between males and females, but also among males, and among females respectively. The quality of social capital is also different among the different social groups due to their different socio-political backgrounds.

Secondly, the socio-political transformation in the region through the CIG approach is weak even though change has begun. Therefore, our hypothesis 2 "CIG is an effective approach to empower women through socio-political transformation (H2)" has been insufficiently verified. The formation of groups in rural communities by the government and non-government organizations are not new to the culture of Bangladesh (DAE 2016). Building self-reliant farmer's groups for increasing access to new knowledge and technology, and local institutions could be an effective approach to empower women in poverty reduction and disaster risk governance if it creates social innovation for socio-political change, especially in entrepreneurship development. However, the issues of poverty and gender discrimination are complex and strongly route-dependent in the region.

Thirdly, according to the above-mentioned SEM analytical results, our hypothesis 3 "Farmers' performance in disaster risk management at community level is determined by their socio-economic attributes, social capital, and access to local institutions (H3)" has been verified. The results also clarified that gender and poverty are important factors for the farmer's performance in disaster risk management in the study area. Furthermore, this research presented important evidence that there are multi-causal relationships among the factors affecting farmer's performance in DRM at the community level. For instance, disaster countermeasures in rural Bangladesh are seen as one of the solutions to break the vicious circle of poverty in rural areas. However, it is necessary to recognize the cause of poverty is strongly route-dependent, such as the existence of severe gender discrimination (Kabeer 2011) and a low trust to government officials. The findings proved that farmer's performance in DRM in the study area can be significantly improved through the elimination of gender discrimination and the accumulation of social capital. In other words, it is necessary to recognize that strengthening the current DRM policies without paying attention to these multi-causal relations may decrease farmers' performances in DRM, and the causes of poverty in rural areas may be hidden.

Therefore, our policy implication suggests that the government should pay more attention to social innovation and entrepreneurship for socio-political transformation through the collective cognitive change of the society along with economic growth. The regional development problems must be considered through a different lens instead of trying to fix the visible signs of poverty and investing in a particular type of innovation (market-creating innovation), which is considered as a catalyst and foundation for creating sustained economic development (Christensen et al. 2019). Empower women in particular by providing access to land and household resources, official licenses to operate farm machinery and creating a safe environment for them to work outside. Finally, introduce financial cooperatives in rural areas for local entrepreneurs to get access to finance through structural change in the financial system. Otherwise, economic growth without socio-political transformation may create new conflicts in the region.

However, appropriate measurement of the dynamism of regional entrepreneurship such as REDI (Regional Entrepreneurship and Development Index) analysis (EU 2014) is required to draw more effective policy implications for enhancing sustainable regional development in the *haor* region of Bangladesh. This will be our next research agenda.

References

Christensen CM, Ojomo E, Dillon K (2019) The prosperity paradox: How innovation can lift nations out of poverty. Harper Business, New York

DAE (2016) Agricultural extension manual. Department of Agricultural Extension, Ministry of Agriculture, Government of the People's Republic of Bangladesh, Dhaka (in Bengali)

EU (2014) REDI: The regional entrepreneurship and development index-measuring regional entrepreneurship. Final report. Publication office of the European Union, Luxembour

Kabeer N (2011) Between affiliation and autonomy: Navigating pathways of women's empowerment and gender justice in rural Bangladesh. Dev Chang 42(2):499–528